Climate Change
Policy after Kyoto

Contents

Foreword

In its core mission to sponsor research on the key public policy issues of our time, the Brookings Institution is especially on the lookout for ways to contribute to the understanding of global climate change and the challenge it poses to our nation and the world. Thanks to generous funding by one of our trustees, Stephen M. Wolf, we are able to launch this book, the first publication of our Energy and Environment program.

Warwick McKibbin and Peter Wilcoxen have been conducting a long-term research project on climate change as nonresident senior fellows in our Economic Studies program, and this book is the result. Warwick is a professor of international economics at the Australian National University, and Peter is an associate professor of economics at the University of Texas at Austin.

Climate change policy has been a subject of intense international debate since the opening of the United Nations Earth Summit in Rio de Janeiro more than a decade ago. The intensity of the debate, however, has not been matched by action. Over the last few years attention has been focused on the Kyoto Protocol, an ambitious agreement that would in theory sharply reduce emissions of carbon dioxide and other greenhouse gases in industrial countries. However, the protocol has yet to enter into force and has been spurned by the United States and developing countries. Should it enter into force in the near future, it will be because emissions targets have been substantially relaxed as a short-term political compromise. Meanwhile, emissions of carbon dioxide and other greenhouse gases continue to grow unchecked, and they will continue to do so if political compromises relaxing targets are all that keep the protocol alive.

In the pages that follow, Warwick and Peter argue that the climate policy stalemate results directly from flaws in the design of the Kyoto Protocol. In their view, the root of the problem is the protocol's focus on establishing targets and timetables for reductions in greenhouse gas emissions. The United States, for example, would be required to reduce its average annual emissions over the period 2008–12 by 7 percent of the amount it emitted in 1990. The authors point out that in order to ratify the treaty, a country must implicitly agree to achieve its emissions target regardless of the cost of doing so. Given the tremendous uncertainties pervading every aspect of climate change, the governments of countries having large greenhouse gas emissions—such as the United States—are unlikely ever to be willing to undertake such an open-ended commitment.

Warwick and Peter carry their critique further to argue that the Kyoto Protocol's emphasis on international emissions trading is politically unrealistic; that the agreement has no credible mechanism for monitoring participants and enforcing compliance; and that it will be particularly vulnerable to collapse if a major country was to withdraw. They devote the remainder of the book to presenting an alternative policy that would address the protocol's shortcomings. They argue that their approach is more appropriate given the uncertainties surrounding climate change and that it demands far less surrender of sovereignty from participating countries. Perhaps most important, they show that their approach would create powerful constituencies with vested interests in the long-term success of the policy. Since a climate policy and the institutions sustaining it will need to remain in effect indefinitely, creating such interest groups will be essential. Brookings is proud to be part of this effort.

STROBE TALBOTT
President, Brookings Institution

Washington, D.C.
October 2002

Acknowledgments

W e are deeply grateful to Brookings trustee Stephen M. Wolf, whose financial support made this book possible. It has been wonderful to have the opportunity to examine the problems of and prospects for climate change policy in more depth than would be possible in journal articles and op-ed pieces. In addition, McKibbin would like to thank the Australian National University and Jan and Patrick Davies for supporting this research.

We are grateful also to Robert Litan and Henry Aaron, the current and former directors of the Brookings Economic Studies program, whose support and encouragement have been outstanding. In addition, we received many helpful comments on this manuscript and a related paper we wrote for the *Journal of Economic Perspectives* from Robert Crandall, J. Bradford De Long, Paul Portney, Timothy Taylor, Michael Waldman, and an anonymous referee. Our work has also been improved by many insightful comments we have received from other colleagues over the years, especially Philip Bagnoli, Ralph Bryant, Jae Edmonds, Ross Garnaut, Lawrence Goulder, Raymond Kopp, Alan Manne, Richard Morgenstern, William Nordhaus, William Pizer, Richard Richels, Robert Shackleton, Robert Stavins, and John Weyant.

We would like to acknowledge our debt to Linda Gianessi, whose excellent—and unfailingly cheerful—help with administrative matters gave us the luxury of focusing on our research. We also appreciate the assistance of Ev Taylor, Theo Merkle, Eileen Robinson, Jennifer Derstine, and Deborah Washington, each of whom helped with many logistical and administrative matters. In addition, we would like to thank several people for their help with the production of this book: Janet Walker, managing editor of the Brookings Press, who supervised the publication process; Eileen Hughes, whose excellent editorial work significantly improved the manuscript; Catherine Theohary, who verified the manuscript; and Julia Petrakis, who prepared the index. We thank, too, Susan Woollen, Brookings art coordinator, for her patient and expert supervision of the cover design process, and Becky Clark, marketing director, for her work to disseminate the findings of this project.

Finally, we are especially grateful to our wives, Jennie McKibbin and Paula Wilcoxen, and our children—Rebecca, Patricia, and Jordan McKibbin and Kaitlin and Stephanie Wilcoxen—whose understanding and encouragement were unwavering. This book would not have been possible without their support.

WARWICK J. MCKIBBIN
PETER J. WILCOXEN

Climate Change
Policy after Kyoto

1 A Realistic Approach to Climate Change

A t the heart of the debate on climate change are two key facts. The first is the familiar and undisputed observation that human activity is rapidly increasing the concentration of greenhouse gases in the atmosphere. Each year, worldwide fossil fuel use adds about 6 billion metric tons of carbon to the atmosphere, and the concentration of carbon dioxide is now about 30 percent higher than it was at the dawn of the Industrial Revolution.

The second fact is equally important but far more subtle: no one fully understands how the climate will respond. The increase in greenhouse gases could lead to a sharp rise in global temperatures, with severe consequences for ecosystems and human societies. On the other hand, it is possible that the rise in temperature could be modest, easy to mitigate, and far in the future. The most likely outcome lies somewhere between the two, but the intrinsic complexity of the climate makes it impossible to know precisely what will happen with any degree of confidence.

That uncertainty has sharply polarized public debate. On one side are those who point to the possibly disastrous consequences of climate change and argue that emissions must be reduced sharply to lower the risk of catastrophe. A typical argument from this group might be that fossil fuel use should be reduced in order to lower the risk of disintegration of the West Antarctic ice sheet, an event that would raise global sea levels by 3 meters. On the other side are those who point to the possibly small probability of a disaster and argue that there are better uses of society's resources than reducing an already small risk even further. A typical argument from this group might be that money would be better spent by reducing conventional pollution, investing in new technology, alleviating poverty, and raising educational standards than by reducing greenhouse gas emissions.

There are elements of truth in both positions, but neither is an appropriate response to climate change. An objective reading of current scientific literature indicates that a moderate effort should be made to slow the growth of greenhouse gas emissions. Taking some sort of action is warranted: although climatologists disagree about the timing and magnitude of warming, no one seriously suggests that mankind can continue to add increasing amounts of carbon dioxide to the atmosphere every year without any adverse consequences. Moreover, climate change is essentially irreversible, so it makes sense to avoid causing more of it than necessary, at least until the potential risks are better understood. At the same time, current evidence does not justify a draconian cut in emissions: the cost would be enormous and the environmental benefits might be small. It is easily possible that the resources needed for a sharp reduction would be better spent on more immediate social problems. As a matter of common sense, therefore, the right approach must lie between the two extremes: the policy should provide incentives to reduce greenhouse gas emissions but avoid imposing unreasonably large costs.

To date, however, common sense has played little role in the debate over climate change policy. Instead, polarization of the participants has led to impractical proposals and a decade of policy deadlock. In 1992, the United Nations Earth Summit in Rio de Janeiro produced a landmark treaty on climate change that proposed stabilizing greenhouse gas concentrations in the atmosphere. By focusing on stabilization, the treaty implicitly adopted the position that the risks posed by climate change require emissions to be reduced no matter what the cost. The agreement, ratified by more than 186 countries, including the United States,

Climate Change Policy after Kyoto

Blueprint for a Realistic Approach

Warwick J. McKibbin
Peter J. Wilcoxen

BROOKINGS INSTITUTION PRESS
Washington, D.C.

Library of Congress Cataloging-in-Publication data

McKibbin, Warwick J., 1957–
    Climate change policy after Kyoto : a blueprint for a realistic
approach / Warwick J. McKibbin and Peter J. Wilcoxen.
        p. cm.
    Includes bibliographical references and index.
    ISBN 0-8157-0608-1 (cloth : alk. paper) —
    ISBN 0-8157-0607-3 (pbk. : alk. paper)
    1. Climatic changes—Government policy. 2. Environmental policy.
I. Wilcoxen, Peter J. II. Title.

QC981.8.C5 M39 2002
363.738'746—dc21   2002151486

9 8 7 6 5 4 3 2 1

The paper used in this publication meets minimum requirements of the
American National Standard for Information Sciences—Permanence of Paper for
Printed Library Materials: ANSI Z39.48-1992.

Typeset in Sabon

Composition by Cynthia Stock
Silver Spring, Maryland

Printed by R. R. Donnelley and Sons
Harrisonburg, Virginia

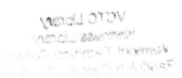

prompted numerous subsequent rounds of climate negotiations aimed at rolling back emissions from industrialized countries to the levels that prevailed in 1990. In the end, however, it has had virtually no effect on greenhouse gas emissions; it has not even produced a detectable slowing in the rate of emissions growth. The treaty's implementing agreement, the 1997 Kyoto Protocol, has been rejected by the United States and spurned by developing countries. Ten years of negotiations have produced a policy that is very strict in principle but completely ineffective in practice.

International negotiations have failed because they have been predicated on the assumption that climate change must be stopped at any cost. However, a climate policy that does not take costs into consideration will never be ratified by the U.S. Senate, and it will be rejected by many other governments as well. One can see why by anticipating the questions that prudent senators will ask when a climate change treaty eventually comes up for ratification. The first question will be, in essence, "Will the treaty's benefits exceed its costs?" Given current scientific understanding of the climate, the answer would have to be "It is impossible to tell." That response alone would not condemn the treaty; political decisions often involve uncertainty. However, it would lead to a second question: "Do we know at least that the cost will not be excessive?" It is this question that would be fatal for a treaty based on the assumption that emissions must be reduced at any cost. The answer would have to be "No," and prudent legislators would have little alternative but to reject the treaty. Any other decision would be irresponsible: it would commit the country to spending a potentially large amount of its resources on a policy that might have very little benefit.

To break the policy deadlock, climate negotiations must be redirected toward a more practical and realistic approach. In order to succeed, the alternative policy must reflect the deep uncertainties in predicting climate change by taking a moderate approach: providing incentives to reduce greenhouse gas emissions but avoiding unreasonably large costs. In addition, the policy's political prospects will be substantially better if it does not require large transfers of wealth or the surrender of a significant degree of national sovereignty. Finally, the system will need to remain in effect for many years, so it must be designed to allow new countries to enter with minimum disruption and to survive the exit of some of its participants.

Neither of the standard market-based economic policy instruments satisfies all of these criteria. An ordinary tradable permit system would require participants to achieve a rigid emissions target regardless of cost. An emissions tax, on the other hand, would involve huge transfers of wealth and would be politically unrealistic. However, a hybrid policy combining the best features of the two would be an efficient and practical approach.[1]

One such hybrid policy would combine a fixed number of tradable, long-term emissions permits with an elastic supply of short-term permits, good only for one year. Each country participating in the policy would be allowed to distribute a specified number of long-term emissions permits, referred to as perpetual permits. The number of perpetual permits would be set by international negotiation; one possibility is that each country would be allowed to issue a quantity equal to the amount of the country's 1990 emissions. The permits could be bought, sold, or leased without restriction, and each one would allow the holder to emit one ton of carbon per year. The permits could be given away, auctioned, or distributed in any other way that the government of each country saw fit. Once distributed, the permits could be traded among firms or bought and retired by environmental groups. In addition, each government would be allowed to sell additional annual permits for a specified fee, say for $10 per ton of carbon.[2] Other things being equal, the price of a $22 ton of coal would rise by about 30 percent and the price of a $20 barrel of oil would rise by 7 percent. Firms within a country would be required to have a total number of emissions permits, in any combination of perpetual and annual permits, equal to the amount of emissions they produce in a year.

1. The economic theory behind hybrid regulatory policies was developed by Roberts and Spence (1976). A hybrid approach to climate change was first proposed by McKibbin and Wilcoxen (1997a) and subsequently has been endorsed or promoted by a range of authors and institutions. Examples include Kopp, Morgenstern, and Pizer (1997); Kopp and others (1999); Aldy, Orszag, and Stiglitz (2001); and Victor (2001).

2. This figure can be translated into more familiar terms using the carbon content of different fuels. A ton of coal contains 0.65 tons of carbon, so a permit fee of $10 per ton of carbon would add $6.50 to the price of a ton of coal. Similarly, a barrel of crude oil contains about 0.14 tons of carbon, so the fee would add $1.40 to the price of a barrel.

The single most important feature of the hybrid is that it establishes an upper limit on the cost of compliance. No firm would have to pay more than $10 per ton to reduce its emissions because it could always buy an annual emissions permit instead. As a result, the answer to the question "Do we know at least that the cost will not be excessive?" would be "Yes." Adopting the hybrid, in other words, does not require a country to make an open-ended commitment to reduce its emissions regardless of cost.

The hybrid policy would have many other desirable attributes as well, which are summarized briefly in the box on page 6 and discussed in detail in chapters 5 and 7. Among the most important is that the policy would be very stable with respect to changes in the mix of participating countries. Because permit markets in different countries would be separate, linked only by the common price of an annual emissions permit, one country's entry into or exit from the system would have no effect on the price of permits circulating in other countries. In contrast, a change in the list of countries participating in the Kyoto Protocol would cause windfall gains or losses to ripple through permit markets around the world.

Moving from the current deadlock to the hybrid approach will require leadership. The United States could jump-start the process by adopting a modified form of the hybrid policy unilaterally. The government could immediately distribute a quantity of perpetual emissions permits equal to the U.S. commitment under the Kyoto Protocol, but with one important caveat: firms would *not* be required to hold emissions permits unless the United States were to ratify an international agreement on climate change. Essentially, the government would distribute contingent property rights for greenhouse gas emissions. Such a step would be bold, but it also would be in the self-interest of the United States. Rapidly rising world carbon dioxide emissions make it inevitable that some sort of climate policy eventually will be adopted. Issuing property rights as soon as possible would help the economy adapt because it would allow financial markets to help manage the risks of climate policy. For example, a firm worried that it would be unable to comply with future climate regulations could reduce its risk by buying extra permits, or even options on extra permits, as a hedge. A firm able to reduce its emissions at low cost could sell permits (or options) now.

A Hybrid Policy for Climate Change

In its basic form, the hybrid policy allows each participating country to issue two kinds of emissions permits: perpetual permits, which entitle the owner of a permit to emit one metric ton of carbon every year forever; and annual permits, which allow one ton of carbon to be emitted in a single, specified year. Key features of the policy are listed below.

Perpetual permits
    Quantity restricted to a specified fraction of 1990 emissions, for example, 95 percent.
    Distributed once, when the policy is first enacted.
    Can be bought, sold, or leased within the country of issue without restriction.
    Can be used only in the country of issue; no international trading.
    Price set by financial markets.

Annual permits
    Sold for a stipulated price, for example, $10 per ton of carbon.
    No limit on the quantity sold.

Because the hybrid policy has two kinds of permits, it is more complicated than a simple permit system. However, it has all of the strengths of a traditional permit system and additional advantages as well. It performs especially well in comparison with the Kyoto Protocol.

Strengths shared with a conventional permit system
    Reduces emissions in a cost-effective manner.
    Perpetual permits avoid the huge transfers of wealth that would occur under a tax.
    Allows historical emissions to be grandfathered, providing transition relief.
    Has built-in incentives for monitoring and enforcement.

Additional advantages compared with the Kyoto Protocol
    Existence of annual permits provides an upper bound on compliance costs.
    Trading is national rather than international.
    Does not rely on large international transfers of wealth.
    Annual permits generate revenue, providing an incentive for enforcement.
    More credible; more likely to be enforced into the future.
    Relatively easy to modify as information arrives.
    Easily adapted to provide incentives for early action.
    Easy to add countries over time; does not require renegotiation of treaty.
    Entry or exit of countries does not affect other permit owners.

Other notable feature compared with the Kyoto Protocol
    Does not guarantee any particular cut in emissions: if abatement costs are high, the overall reduction will be less than under Kyoto.

Of course, pricing these permits would present a short-run challenge for financial markets, since it is uncertain when carbon emissions will be regulated. But financial markets confront this kind of problem every day. Within a very short time, an active market would develop with prices that reflected both the likelihood of a policy taking effect and its probable stringency. Indeed, active markets already have been formed for trading privately created emissions permits.[3]

Overall, a hybrid climate change policy has much to offer. It is flexible enough to deal with the enormous uncertainties regarding climate change. It provides individual governments with an instrument to limit and channel the distributional effects of the policy on different groups within their own countries, reducing the obstacles to ratification. Moreover, we will show that it creates incentives for governments to monitor and enforce the policy within their own borders. It is a practical policy that would reduce greenhouse gases in a cost-effective manner.

The remainder of the book presents the hybrid policy in more detail. Chapters 2 through 4 provide background information on the climate change debate. Chapter 2 summarizes the current state of knowledge about climate change—the science of climate change, the impacts, and the costs of reducing it. Chapter 3 outlines the history of climate change negotiations since the adoption in 1992 of the United Nations Framework Convention on Climate Change (UNFCCC), discussing the 1997 Kyoto Protocol and explaining how it has evolved. Chapter 4 examines the Kyoto Protocol in detail to show that its approach to climate change policy is fundamentally incorrect.

The remaining chapters, 5 through 8, focus on our proposal for a hybrid policy. Chapter 5 presents the economic theory behind it; chapter 6 discusses the practical details of its implementation, including several extensions and refinements that could be adopted; and chapter 7 evaluates its strengths and weaknesses relative to those of several benchmark policies. Finally, chapter 8 outlines a process that could be used to break the current deadlock in international negotiations.

3. Rosenzweig and others (2002).

2 Science and Uncertainty

A t the heart of the climate change debate are two undisputed facts. The first is that certain gases in the atmosphere allow visible and ultraviolet light to pass but absorb infrared radiation. The most famous of these gases is carbon dioxide (CO_2), but water vapor, methane (CH_4), nitrous oxide (N_2O), halocarbons (which include chlorofluorocarbons), and various other gases have the same property. In 1895, Svante Arrhenius, a Swedish chemist, showed that the presence of carbon dioxide in the atmosphere raises the earth's surface temperature substantially.[1] Energy from the sun, in the form of visible and ultraviolet light, passes through the carbon dioxide unimpeded and is absorbed by

---

1. Arrhenius (1896) calculated that removing all carbon dioxide from the atmosphere would lower global temperatures by about 31°C (56°F). The direct effect of removing the carbon dioxide would be to lower temperatures by 21°C (38°F). In addition, the cooler air would hold less water vapor, lowering temperatures by another 10°C (18°F). For comparison, the actual global average temperature is about 14°C (57°F), and a change of this magnitude would give Los Angeles a climate roughly like that of Nome, Alaska.

objects on the ground. As the objects become warm, they release the energy as infrared radiation. If there were no carbon dioxide in the atmosphere, most of the infrared energy would escape back into space. The carbon dioxide, however, absorbs the infrared energy and reradiates it back toward the surface, thus raising global temperatures. This mechanism is now known as the "greenhouse effect" because it traps energy near the earth's surface in a manner somewhat analogous to the way that glass keeps the air inside a greenhouse warm. Carbon dioxide and other gases contributing to this effect are called "greenhouse gases."

The second undisputed fact is that the concentration of many greenhouse gases has been increasing rapidly because of human activity. Since the Industrial Revolution, deforestation and the use of fossil fuels has raised the level of carbon dioxide in the atmosphere by about 30 percent.[2] Emissions rose sharply in the latter half of the twentieth century, as shown in figure 2-1.

Methane is a much smaller fraction of the atmosphere, but its concentration has risen even more rapidly: it is now about 150 percent above preindustrial levels.[3] Global methane emissions since the middle of the nineteenth century are shown in figure 2-2. In addition, the concentration of nitrous oxide has risen by 17 percent, and all halocarbons in the atmosphere are the result of human activity.

Greenhouse gases differ from one another in the amount of energy they trap per kilogram, which is called the "radiative forcing" of the gas and is measured in watts per square meter of the earth's surface (Wm^{-2}). Methane, for example, traps substantially more energy than an equal amount of carbon dioxide. Halocarbons, including chlorofluorocarbons, trap even more. As a result, methane and halocarbon emissions have a much larger potential effect on the climate than would be expected from the quantity of their emissions alone. Table 2-1 shows the radiative

2. Ice cores and other evidence show that the atmospheric concentration of carbon dioxide was about 280 ± 10 parts per million (ppm) for several thousand years before the beginning of the Industrial Revolution. By 1998, the concentration had risen to 365 ppm.

3. The concentration of methane has risen from about 700 parts per billion (PPB) before 1750 to 1,745 PPB in 1998. The sources and sinks of atmospheric methane are less well understood than those of carbon dioxide. However, significant anthropogenic sources include agriculture, natural gas production, and landfills.

Figure 2-1. *Global Carbon Dioxide Emissions, 1751–1998*

Millions of metric tons

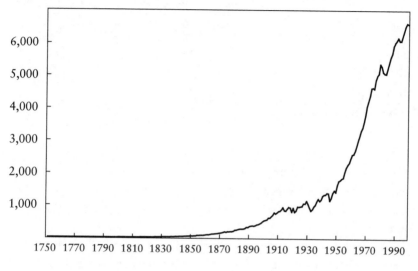

Source: Marland, Boden, and Andres (1998).

Figure 2-2. *Global Methane Emissions, 1860–1994*

Millions of metric tons

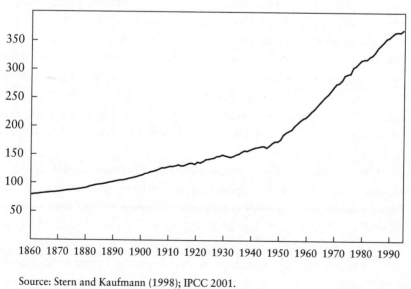

Source: Stern and Kaufmann (1998); IPCC 2001.

Table 2-1. *Radiative Forcings of Current Greenhouse Gas Concentrations*

Gas	Radiative forcing Wm^{-2}	Percent of total
Carbon dioxide	1.46	60
Methane	0.48	20
Nitrous oxide	0.15	6
Halocarbons	0.34	14
Total	2.43	100

Source: IPCC (2001a).

forcings associated with the increase in greenhouse gas concentrations from preindustrial levels. Carbon dioxide is by far the largest greenhouse gas in terms of emissions, but it accounts for only about 60 percent of the potential problem.

The length of time that a kilogram of gas typically remains in the atmosphere before being removed through a chemical reaction is known as the "atmospheric lifetime" of the gas. The atmospheric lifetime of carbon dioxide can be as long as 200 years. Methane, on the other hand, has a lifetime of only twelve years. The lifetimes of some halocarbons are enormous: perfluoromethane (CF_4) remains in the atmosphere for 50,000 years or more. The total amount of energy trapped by a kilogram of a greenhouse gas over a specified number of years relative to the amount that would be trapped by a kilogram of carbon dioxide over the same period is called the "global warming potential" (GWP) of the gas. Table 2-2 shows the global warming potential for carbon dioxide, methane, nitrous oxide, and selected halocarbons over a 100-year period. HFC-23 is included because it is a common substitute for chlorofluorocarbons, which are being eliminated under the Montreal protocol on substances that deplete the ozone layer.

Most greenhouse gas emissions originate in industrialized countries, although emissions in developing countries have been rising rapidly in recent years. In 1999, industrial countries were responsible for 69 percent of world carbon dioxide emissions: 25 percent came from the United States, 31 percent from other Organisation for Economic Co-operation and Development (OECD) countries, and 13 percent from countries in

Table 2-2. *Global Warming Potential for Selected*
Greenhouse Gases

Gas	Atmospheric lifetime (years)	GWP relative to CO_2 over 100 years
Carbon dioxide	50–200	1
Methane	12	23
Nitrous oxide	114	296
HFC-23 (CHF_3)	260	12,000
Perfluoromethane (CF_4)	50,000	5,700
Sulfur hexafluoride (SF_6)	3,200	22,200

Source: IPCC (2001a).

eastern Europe and the former Soviet Union. More figures are presented in table 2-3, which gives carbon dioxide emissions in 1992 and 1999 for selected countries and regions. Between 1992 and 1999, emissions increased about 10 to 12 percent in many OECD countries, although sharp drops in emissions in Germany and the United Kingdom kept the increase in total OECD emissions to 9 percent. Emissions from eastern Europe and the former Soviet Union declined by 30 percent as those countries underwent sharp economic decline or restructuring. Emissions in developing countries, on the other hand, grew by 22 percent.

Emissions of methane are somewhat less concentrated in industrial countries. Figure 2-3 shows methane emissions in 1990 from selected countries and regions. Emissions are broken down by source: fossil fuel combustion (mostly related to residential energy use), fossil fuel production (mostly related to coal mining and gas pipelines), biofuels, industrial processes, and land use and waste treatment (mostly related to landfills, rice paddies, and byproducts of digestion by ruminant animals).

Emissions of nitrous oxide also are less concentrated in industrial countries. Figure 2-4 shows nitrous oxide emissions in 1990 from selected countries and regions. Emissions are broken down by source: fossil fuel use, biofuels, industrial processes, and land use and waste treatment (mostly related to fertilizers and animal waste).

Because the United States is the single largest source of carbon dioxide emissions and a major emitter of other greenhouse gases, it is useful to examine its emissions in more detail. Table 2-4 shows U.S. emissions in

Table 2-3. *Recent Carbon Emissions by Region and Country*

Region	Carbon (mmt) 1992	1999	Percent of 1999 total	Percent increase from 1992 to 1999
North America	1,577	1,772	29	12
Canada	124	151	2	22
Mexico	86	101	2	17
United States	1,366	1,520	25	11
Central and South America	200	267	4	34
Brazil	64	89	1	39
Other	136	178	3	31
Western Europe	964	1015	17	5
Germany	241	230	4	-5
UK	157	152	2	-3
Other	566	633	10	12
Eastern Europe and the former Soviet Union	1,123	789	13	-30
Russia	574	400	7	-30
Other	549	389	6	-29
Middle East	222	287	5	29
Africa	208	237	4	14
Far East and Oceania	1,537	1,776	29	16
Australia	77	94	2	22
China	668	669	11	0
India	176	243	4	38
Japan	286	307	5	7
Other	330	463	8	40
World total	5,831	6,143	100	5
OECD countries	3,067	3,326	56	9
Industrial countries[a]	4,259	4,222	69	-1
Nonindustrial countries	1,572	1,921	31	22

Source: Energy Information Administration (1999). The figures in this table include carbon resulting from natural gas flaring, which is not included in previous tables.

a. Industrial countries include OECD countries plus Eastern Europe and countries of the former Soviet Union.

1999. The emission of carbon dioxide dwarfs that of the other gases in terms of sheer kilograms: nearly 5,600 million metric tons of carbon dioxide compared with only about 30 million metric tons of other gases. However, the other gases have higher global warming potential, so a better comparison is found in the third column, which shows the amount of carbon that would generate the same long-term atmospheric heating

Figure 2-3. *Emissions of Methane in 1990, by Country and Source*

Millions of metric tons

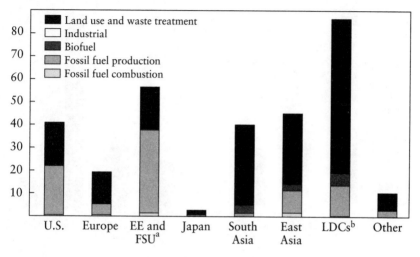

Source: Olivier and Berdowski (2001).
a. Eastern Europe and the former Soviet Union.
b. Less developed countries.

effect.[4] Together, gases other than carbon dioxide account for about 17 percent of U.S. emissions on a carbon-equivalent basis.

U.S. carbon emissions are broken down by fuel and sector of end use in table 2-5. About one-third of U.S. emissions is from petroleum fuels used in transportation; another third is from coal used to power electric utilities; and the remaining third is generated by other industrial, commercial, and residential activities.

U.S. emissions of other greenhouse gases are less significant in global warming. In 1999, the U.S. emitted 28.8 million metric tons of methane, which has an effect on global temperatures equivalent to that of 165 million metric tons of carbon (about 11 percent of U.S. carbon emissions).[5] Major U.S. methane sources are shown in table 2-6.

4. A ton of carbon dioxide contains 0.27 tons of carbon, which is the reason why the first two numbers in the carbon dioxide row differ.
5. Energy Information Agency (2000b).

Figure 2-4. *Emissions of Nitrous Oxide in 1990,*
by Country and Source

Thousands of metric tons

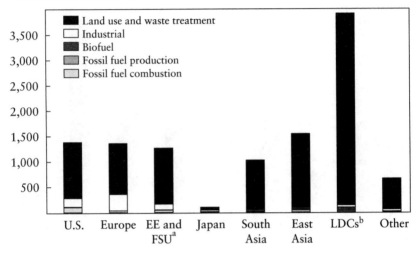

Source: Olivier and Berdowski (2001).
a. Eastern Europe and the former Soviet Union.
b. Less developed countries.

Table 2-4. *U.S. Greenhouse Gas Emissions in 1999*

Gas	Emissions (mmt of gas)	Carbon equivalent (mmt carbon)	Carbon equivalent (%)
Carbon dioxide	5,598.2	1,527	83
Methane	28.8	165	9
Nitrous oxide	1.2	103	6
Halocarbons[a]	< 0.05	38	2
Total		1,833	100

Source: Energy Information Administration (2000b).
a. Halocarbons are carbon compounds containing fluorine, chlorine, bromine, or iodine. Notable halocarbons associated with global warming include chlorofluorocarbons (CFCs); hydrofluorocarbons (HFCs); perfluorocarbons (PFCs), including CF_4 and C_2F_6; and sulfur hexafluoride (SF_6). CFCs also destroy stratospheric ozone and are controlled under the Montreal Protocol; HFCs do not destroy the ozone layer and are some of the leading substitutes for CFCs.

Table 2-5. *U.S. 1999 Carbon Emissions by Sector and Fuel*[a]

| | By primary fuel | | | | | By end use | | |
| | Fuel | | | | | Electric utility | | |
Source	Petroleum	Natural gas	Coal	Total emissions	Percent of emissions	emissions by sector of end use	Total emissions	Percent of emissions
Residential	26	70	1	97	6	193	290	19
Commercial	14	45	2	61	4	183	244	16
Industrial	104	142	56	302	20	180	481	32
Transportation	486	10	0	496	33	1	496	33
Electric utilities	20	46	490	556	37			
Total	650	312	549	1,512		557	1,511	
Percent	43	21	36					

Source: Energy Information Agency (2000a).
a. Million metric tons of carbon unless otherwise noted.

U.S. emissions of nitrous oxide were 1.2 million metric tons in 1999, which is equivalent to about 103 million metric tons of carbon.[6] About three-quarters of that comes from agriculture, with the remainder coming from automobile exhaust and industrial sources. Emissions of halocarbons are very small, but many of the gases have a long atmospheric lifetime and high global warming potential. As a group they account for about 2 percent of U.S. greenhouse gas emissions.

What Might Happen to Future Emissions without a Policy?

As indicated by the figures above, world greenhouse gas emissions have been growing rapidly. In the absence of an effective climate change policy, they will continue to do so for decades. However, it is very difficult to predict the quantity of future emissions with any precision. The sources of the difficulty fall into two broad categories: uncertainty about the appropriate way to model the world economy and uncertainty about the future values of variables that affect the model but are determined outside of it. The first category can be referred to as "structural uncertainty"; an

6. Energy Information Agency (2000b).

Table 2-6. *Major Sources of U.S. Methane Emissions in 1999*

Source	Emissions (mmt)	Percent of total
Energy		
Crude oil and gas extraction	7.03	24
Coal mining	2.88	10
Waste management		
Landfills	8.94	31
Agriculture		
Enteric fermentation[a]	5.40	19
Animal waste	3.03	11
All other sources	1.48	5
Total	28.76	100

Source: Energy Information Agency (2000a).

a. Ruminants such as cows, sheep, goats, and camels produce methane as a byproduct of digestion.

example is the elasticity of demand for gasoline, which reflects people's willingness to reduce gasoline consumption when prices rise. The second category can be called "scenario uncertainty"; examples include the pricing behavior of the Organization of the Petroleum Exporting Countries (OPEC), the rate of population growth, and the degree of social and economic integration among countries.[7]

The most extensive and well-known attempts to characterize future greenhouse gas emissions have been a series of reports by the Intergovernmental Panel on Climate Change (IPCC).[8] The most recent of these (IPCC 2000) assesses the importance of structural uncertainties by using six well-known energy-economy models to forecast greenhouse gas emissions through 2100 for specified scenarios of forcing variables.[9] To mea-

7. In economic terminology, "scenario uncertainty" is uncertainty about the future values of a model's exogenous variables.

8. The IPCC was established in 1988 by the World Meteorological Organization and the United Nations Environment Programme to assess the state of scientific, technical, and socioeconomic information relevant to climate change. Its function is to review and summarize the scientific literature by bringing together hundreds of scientists from a variety of disciplines and countries. It has produced three major assessments—IPCC 1990, 1995, and 2001a, b, and c—and many shorter reports. See http://www.ipcc.ch/ [August 2002] for more information.

9. The models were AIM, ASF, IMAGE, MARIA, MESSAGE, and MiniCAM.

Figure 2-5. *Emissions of Carbon in the MESSAGE Model, by IPCC Scenario*

Gigatons of carbon

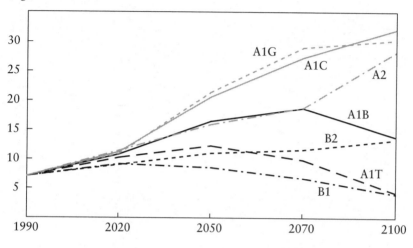

Source: IPCC (2000).

sure the effect of scenario uncertainty, each model was used to calculate emissions under a range of scenarios having standardized values for population growth, technical change, and other variables. Figure 2-5 shows the path of world carbon emissions generated by one of the models, MESSAGE, under seven of the main IPCC scenarios. The differences among scenarios are staggering: emissions in 2050 vary between 8.57 Gt (a 20 percent increase from 1990) and 21.42 Gt (just over a 200 percent increase), and the range in 2100 is much larger. It is important to note that these are not a family of curves around a likely central path: the report explicitly avoids assigning probabilities to any of the trajectories.

The importance of structural uncertainties can be seen by comparing the results for a single scenario. Figure 2-6 shows carbon emissions paths for the six models under the IPCC's A1B scenario. Again, the differences are large: emissions in 2050 vary from 13 Gt to 26.6 Gt across the models, a range of 100 percent.

Other greenhouse gases show similar patterns of uncertainty. Overall, it is likely that emissions will grow substantially in coming decades, but

Figure 2-6. *Emissions of Carbon under IPCC Scenario A1B*

Gigatons of carbon

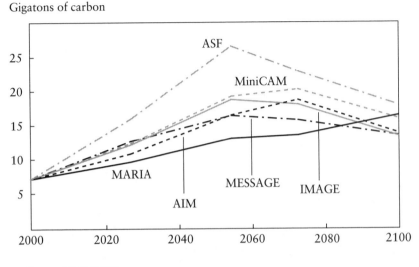

Source: IPCC (2000).

the exact amount is highly uncertain. Moreover, there is no reason to believe that the uncertainties will be reduced very much in the near future.

How Are Global Temperatures Likely to be Affected?

Although it is clear that greenhouse gases trap energy and make the atmosphere warmer and that the concentration of those gases has been increasing, it is not clear what those facts mean for global temperatures. A long list of scientific uncertainties makes it difficult to say precisely how much warming will result from a given increase in greenhouse gas concentrations, when it will occur, or how it will affect different regions and ecosystems.

To calculate the effect of increased greenhouse gas concentrations, climatologists use large computer models of the earth's atmosphere, oceans, and land surface called "general circulation models" (GCMs). GCMs are constructed by dividing the surface of the globe into a horizontal grid, with each grid cell typically 250 kilometers on a side. The atmosphere above each cell is divided into layers about 1 kilometer thick. Where grid cells lie over oceans, the water beneath them is divided into layers 200 to 400 meters thick. Within each cell, physical, chemical, and biological phenomena relevant to climate change are represented by a system of differential equations.

The number of vertical layers and the size of the horizontal grid determine the resolution of the model: more layers and a smaller grid size increase the resolution of the model and enable it to do a better job of capturing small-scale phenomena. That is important because many climate phenomena are very small scale: cloud formations, in particular, are tiny compared with a $250 \times 250 \times 1$ kilometer (62,200-cubic-kilometer) atmospheric cell in a GCM. As resolution increases, however, the computing power needed to solve model increases as well.

Modeling the physical processes going on within each cell is difficult because many of the important phenomena are not fully understood. One challenge for climatologists has been the link between temperature change and atmospheric water vapor. Higher temperatures increase the rate of evaporation and also allow the atmosphere to hold more water vapor. That in turn can lead to positive feedback: because water vapor is itself a greenhouse gas, more water vapor could trap more infrared energy and raise temperatures further. In most GCMs, this effect accounts for about half of the predicted change in temperature resulting from a given change in carbon dioxide concentrations.

However, the role of water vapor in the atmosphere is not completely understood, and only in the last few years has systematic data collection been undertaken. Much of the atmosphere is unsaturated, so an increase in the atmosphere's water vapor capacity does not necessarily translate into an equal increase in actual water vapor. In addition, the effect of temperature on water vapor in the upper troposphere is controversial. Some climatologists argue that increased precipitation could actually reduce water vapor at high altitudes in the tropics.[10]

A closely related uncertainty is that regarding the role of clouds. Clouds reflect incoming radiation, so an increase in cloud cover could tend to

---

10. IPCC (2001a), p. 49. The argument that water vapor could lead to negative feedback is presented by Richard Lindzen and his colleagues; see Hansen (1999) for a discussion of this point.

reduce the greenhouse effect. At the same time, they absorb and reradiate infrared energy and therefore tend to increase the greenhouse effect. Which effect dominates depends heavily on factors that vary from one location to another: the altitude and thickness of the clouds, the amount of water vapor in the atmosphere, and the presence of ice crystals or aerosols (tiny airborne particles or droplets) in the area. Given current knowledge, it is not possible to say for certain whether cloud formations are likely to amplify or attenuate temperature changes from other sources.

Another problem is determining how quickly ocean temperatures will respond to global warming. Water has a high heat capacity, and the volume of sea water is enormous, so the oceans will tend to slow climate change by absorbing excess heat from the atmosphere. This effect delays warming but does not prevent it: eventually the oceans will warm enough to return to thermal equilibrium with the atmosphere. The time required to reach equilibrium depends on many complicated interactions, such as the mixing of different layers of sea water, that are not completely understood and are difficult to model.

Yet another important uncertainty arises because the role of aerosols in the atmosphere is poorly understood. Aerosols originate from a variety of sources: dust storms, volcanoes, fossil fuel combustion, and the burning of forests or other organic material. They reflect a portion of incoming solar radiation and therefore tend to reduce climate change, but they also absorb infrared energy and therefore tend to increase it. Aerosols are poorly understood in a number of respects, but their concentration in the atmosphere seems to be increasing, and their net effect on global temperature seems to be negative.[11] That is particularly true of sulfate aerosols, which arise in part from the sulfur dioxide emitted when sulfur-containing fossil fuels are burned. Before the mid-1990s, GCMs did not include sulfate aerosols. As a result they predicted that past greenhouse gas emissions would result in more warming than has actually been observed. It appears, in other words, that aerosols have partially offset the effects of the increase in carbon dioxide during the last century.

---

11. The Third Assessment Report of the Intergovernmental Panel on Climate Change describes the current level of scientific understanding of aerosols as "low" to "very low" (IPCC, 2001a).

There are a variety of scientifically plausible ways to approach each of these issues, and as a result there are large differences among GCMs. Climatologists compare GCMs by measuring their long-term temperature response to a particular benchmark experiment: a doubling of the atmospheric concentration of carbon dioxide from preindustrial levels. The induced change in mean surface temperature is known as the "climate sensitivity" of the model. The variation in climate sensitivity among GCMs and within a single GCM when its parameters and assumptions are varied is a rough measure of the net size of the uncertainties in climate modeling. Evaluating current GCM results, IPCC (2001a) concludes that the earth's climate sensitivity is "likely" to be somewhere between 1.5°C and 4.5°C, where "likely" is defined to mean having a 66 to 90 percent probability.[12] The chance that the sensitivity is above or below that range is therefore 10 to 34 percent.

Most of the underlying uncertainties are very difficult to resolve, and it is unlikely that the range of possible climate sensitivities will be narrowed much in the coming decades. Using a very simple model with limited data, Arrhenius performed the first climate sensitivity calculations in 1895 and determined that doubling the concentration of carbon dioxide in the atmosphere from preindustrial levels would raise global average temperatures by 4° to 6°C.[13] Although modern methods suggest that the mean increase is smaller, the range of uncertainty has actually grown, from 2° (4° to 6°) to 3° (1.5° to 4.5°), as climatologists have become aware of more complicated and subtle mechanisms at work in the atmosphere, such as the role of aerosols.

Partly because it has proved so difficult to predict the future effect of greenhouse gases on global temperatures, many participants in the

---

12 For consistency, IPCC (2001a) uses the term "very likely" to mean a 90 to 99 percent chance, "likely" to mean a 66 to 90 percent chance, "medium" to mean a 33 to 66 percent chance, and "unlikely" to mean a 10 to 33 percent chance. Events are assigned to these categories based on a rough consensus among climatologists rather than by formal probability analysis.

13. Ice cores and other evidence show that the atmospheric concentration of carbon dioxide was about 280 ± 10 parts per million for several thousand years before the beginning of the Industrial Revolution. Doubling that would be a concentration of 560 ppm. By 1998, the actual concentration of carbon dioxide had risen to 365 ppm, and it is likely that it will reach 560 ppm some time between 2050 and 2100.

Figure 2-7. *Global Temperature Record, Vostok Ice Core Data*

Difference in mean temperature, °C

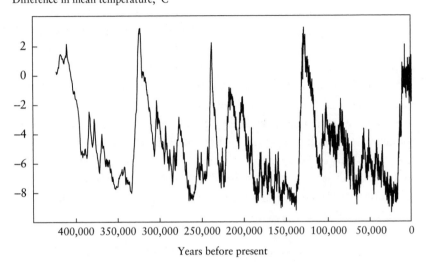

Years before present

Source: Petit and others (2000).

debate—scientists, government officials, environmentalists, and lobby-ists—have been interested in finding out whether past emissions have had much effect. Here, too, certainty is elusive. In spite of articles in the popular press that report every hot summer as evidence of global warming and every cold winter as evidence against it, it is actually quite hard to prove that global warming has begun to happen. Normal variations in global temperatures are large compared with the change that might have been caused by past emissions of greenhouse gases. It is very difficult to tell whether actual increases in temperature are outside the usual range, and thus hard to tell whether greenhouse gases have caused any global warming. Figure 2-7 shows variations in global temperatures over the last 450,000 years, inferred from ice core records from Vostok, Antarctica.

The earth's temperature has varied considerably over time, and the last 10,000 years or so have been warmer than average. Widespread systematic measurements of surface temperatures have been made since the middle of the nineteenth century. However, because over the years tem-

Figure 2-8. *Mean Global Surface Temperature, 1856–1999*[a]

Difference in °C

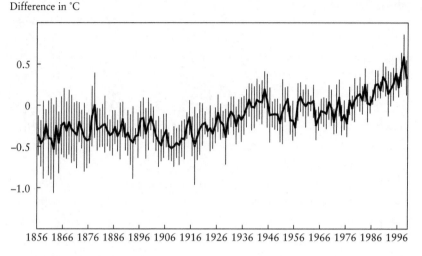

Source: Jones and others (2000).
a. Deviation from 1961–90 mean.

peratures have been measured with different kinds of instruments and at different locations, the data must be analyzed carefully. A simple example of this problem is the urban heat island effect: over time, temperature measurements have been increasingly concentrated in cities, which tend to be warmer than their surroundings. Without correcting for this effect, average temperatures appear to have increased much more than they actually have.

The best estimate of the mean global surface temperature since 1856, after adjusting for known problems in the data, is shown in figure 2-8. Each year's temperature is relative to the mean temperature from 1961 to 1990; the vertical bars show a confidence interval of two standard deviations for each year's mean temperature.[14]

The series appears to have a distinctly upward trend during two periods: from the beginning of the twentieth century until World War II and

14. The confidence intervals were calculated using the standard deviation of the seasonal temperature differentials.

Figure 2-9. *Mean Global Tropopause Temperature, 1958–99*[a]

Difference in °C

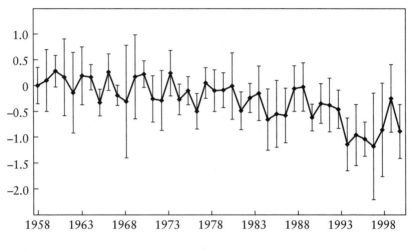

Source: Angell (2000).
a. Deviation from 1958–77 mean.

then again from the late 1970s through the end of the century. More recently, however, observations have been made at higher altitudes using weather balloons (since the late 1950s) and satellites (since 1979). In the part of the atmosphere closest to the earth, the lower troposphere, balloon measurements agree well with surface measurements through the late 1970s, showing warming of about 1°C per decade. After that, however, balloon and satellite observations show substantially less evidence of warming in the lower troposphere than has been observed at the surface. In fact, the data seem to show that the upper layers of the atmosphere have actually cooled in recent years. Figure 2-9 shows mean global temperatures in the tropopause (the layer of the atmosphere from about 10 to 15 kilometers above the surface) as measured by weather balloons; values are relative to the mean temperature from 1958 to 1977.

Surface and tropopause temperatures are compared in figure 2-10, which shows both sets of temperature data for 1958–99 with both normalized to be relative to mean temperatures over 1961–90. The difference between the two series is statistically significant. Moreover, the discrepancy is larger higher in the atmosphere: in the upper troposphere, there

Figure 2-10. *Surface and Tropopause Temperature Trends, 1958–99*[a]

Difference in °C

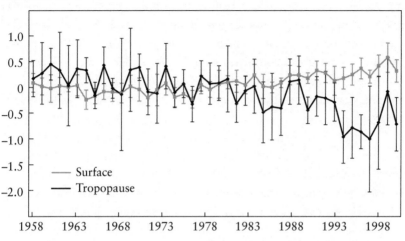

Source: Jones and others (2000) and Angell (2000).
a. Deviation from 1961–90 mean.

has been no trend in temperature since the 1960s, and there has been a clear cooling trend in the stratosphere.

Both IPCC (2001a) and the National Academy of Sciences (2000) have examined the evidence on global temperatures and concluded that in fact mean temperatures have risen. The IPCC, for example, concluded that during the twentieth century global average surface temperatures increased by 0.6° ± 0.2°C. However, the acknowledged inconsistencies in the data make it clear that the conclusion is somewhat uncertain, and climatologists are far from unanimous in their agreement with the IPCC's finding.

The IPCC goes on to conclude that "most of the observed warming over the last 50 years is likely to have been due to the increase in [anthropogenic] greenhouse gas emissions," where "likely" is defined to mean having a 66 to 90 percent probability.[15] This conclusion is suggestive, but

15. Intergovernmental Panel on Climate Change (2001a), "Summary for Policy Makers," pp. 2 and 10. See footnote 12 for details on the terminology IPCC (2001a) uses to describe the likelihood of events.

it is very important to keep in mind that under the usual statistical standards in economics this result would not be significant: it is *not* possible to reject the hypothesis that the warming is *not* anthropogenic. The increase is small, for example, compared with the long-term climate variability shown in the Vostok ice core data. As a result, it is a substantial overstatement of current scientific knowledge to conclude that anthropogenic warming has been detected in historical data. At the same time, however, it is important to remember that all of these measurement problems make it equally difficult to prove that global warming has *not* begun to occur.

In sum, it is impossible to say exactly how much warming has occurred to date or how much will occur in the twenty-first century. Current research summarized in IPCC (2001a) finds that the concentration of carbon dioxide in the atmosphere in 2100 will exceed preindustrial levels by 75 to 350 percent, an enormous range of uncertainty. The source of this uncertainty is essentially economic and very difficult to resolve: predicting the concentration of carbon dioxide in 2100 requires predicting the trajectory of carbon dioxide emissions over the entire preceding century.[16] Other greenhouse gas concentrations will increase as well. Moreover, even if changes in greenhouse gas concentrations could be predicted with complete certainty, there are still large uncertainties in climatology that make it difficult to determine exactly how global temperatures would be affected. Predictions of the global average surface temperature in 2100 show increases of 1.4°C to 5.8°C above the 1990 average. As large as that range is, it still does not include all known uncertainties. Given the complexity of the processes involved, it is not likely that scientists will be able to reduce the uncertainty for decades.

What Else Will Happen to the Climate?

Higher greenhouse gas concentrations will affect the climate in many ways besides simply increasing global average temperatures. General circulation

---

16. The trajectory of carbon emissions will depend heavily on, among other things, population growth, technical change, income growth, and energy prices, none of which is easy to predict over the century.

models show that temperatures in some locations, particularly high-latitude areas in the northern hemisphere, are "very likely" to increase more than the global average, with "very likely" defined by the IPCC to mean having a 90 to 99 percent probability.[17] Other areas, such as the tropics and mid-latitudes in the southern hemisphere, will increase less than average.

GCMs also show that global precipitation will increase, although with substantial variation across regions. Moreover, the increase in precipitation is likely to be associated with an increase in the variability of precipitation from year to year. It is very likely that there will be more intense precipitation over many areas; however, at the same time, the probability of summer droughts over continental areas at mid-latitudes is likely to increase. Precipitation also is likely to become more intense in tropical storms—cyclones, hurricanes, and typhoons—and peak wind speeds in those storms are likely to increase.[18]

Warming is expected to cause sea levels to rise between 9 and 88 centimeters (3.5 inches to 2.9 feet) by 2100. Much of the rise will be due to thermal expansion of the upper layers of water in the oceans, with a smaller but significant contribution from the melting of glaciers. Contrary to science fiction accounts of global warming, the polar ice caps are unlikely to have a major effect on sea level. Warming is likely to reduce the amount of ice in Greenland but to increase it in the Antarctic, which is thought likely to receive an increase in precipitation. There is likely to be a reduction in sea ice in the northern hemisphere, but it will have little effect on the sea level.[19] Two events that would cause a catastrophic rise in sea level—complete melting of the Greenland ice sheet or disintegration of the West Antarctic ice sheet, either of which would raise the sea level by 3 meters—are now thought to be very unlikely in the foreseeable future.[20]

17. See footnote 12 for details on the terminology IPCC (2001a) uses to describe the likelihood of events.

18. The intensity of mid-latitude storms also may be affected, but there is currently little agreement among models about the size of the effect.

19. This is true because floating ice displaces approximately the same amount of liquid water.

20. Climate models indicate that the risk of precipitous melting is very low until at least 2100. Beyond that, the risk would begin to rise, but only if emissions continue to grow unchecked for generations.

Finally, warming may cause changes in key ocean currents. An important example is the Atlantic thermohaline circulation (THC), in which differences in water temperature and salinity drive warm surface water north from the tropics, keeping European winters milder than they would be otherwise. Evidence from ice cores indicates that the Atlantic THC can change very rapidly, but current GCMs do not agree on how it would be affected by global warming.

What Are the Impacts of Climate Change?

Climate change will produce a variety of effects on ecosystems and human activities. However, the effects are less certain than the changes in climate discussed above. They depend on the amount of warming, which is uncertain, but they also involve significant additional uncertainties. This section presents a brief overview of the broad scope of possible effects and the degree of uncertainty associated with each. IPCC (2001b) examines climate change impacts in more detail.

The most obvious impact of higher global temperatures will be to increase energy demand for cooling and reduce demand for heating. However, the net effect on energy demand will vary by region and climate change scenario.

Human health also will be affected. There will be an increase in heat-related and a decrease in cold-related injury and mortality. IPCC concludes with "medium confidence" that a net improvement in mortality will occur in developed countries in temperate regions.[21] It predicts, also

---

21. IPCC (2001b) and IPCC (2001a) use slightly different terminology to quantify uncertainty. When a conclusion is described as holding with "medium confidence," it means essentially that the null hypothesis can be rejected at a level between 33 and 67 percent. Other confidence levels are defined as follows: "very high" means 95 percent or greater; "high" means 67 to 95 percent; "low" means 5 to 33 percent; and "very low" means 5 percent or less. The important difference between this and the scheme used in IPCC (2001a) is at the low end of the scale. Under IPCC (2001a), a "very unlikely" effect has a 1 to 10 percent chance of occurring and a 90 to 99 percent chance of *not* occurring. Under IPCC (2001b), a "very low confidence" conclusion means that the uncertainty is so large that the null hypothesis can be rejected at only the 5 percent level. It does *not* mean that the null hypothesis is true with 95 percent probability.

with medium confidence, an increase in the rate of formation of ground-level ozone, an air pollutant linked to respiratory ailments. In addition, it predicts with medium to high confidence a moderate increase in the population exposed to malaria, dengue fever, and other insect-borne diseases, while expressing medium confidence in the possibility of an increase in the prevalence of water-borne diseases such as cholera.

Higher sea levels would cause a range of problems. Low-lying coastal areas would be inundated (high confidence); the largest areas at risk are river deltas in developing countries. For example, a 45-centimeter rise would inundate about 11 percent of Bangladesh, affecting 5.5 million people; a 100-centimeter rise would inundate almost 21 percent, affecting 14.8 million. Indonesia and Vietnam also would be severely affected, as would a number of small island countries (high confidence). Other coastal areas would be subject to greater damage from erosion; beaches, dunes, and coastal wetlands would be particularly vulnerable. Coastal aquifers throughout the world would be likely to increase in salinity due to saltwater intrusion from higher sea levels (high confidence). Finally, the global population exposed to flooding during coastal storms would increase by 75 to 200 million.

The effects of climate change on agriculture are ambiguous, at least in the aggregate. Many crops in temperate regions would actually benefit from a modest amount of climate change: they would gain more from higher carbon dioxide concentrations than they would be hurt by moderate increases in temperature. However, that benefit would vary strongly by region and crop. Tropical crops would generally be hurt because higher temperatures would lead to increased heat stress. Agriculture in arid regions would be hurt: most GCMs predict that climate change will result in a net decrease in available water in those areas. Overall, IPCC (2001b) concludes that there will be a small positive effect on agriculture in developed countries and a small negative effect in developing countries. This conclusion, however, is highly uncertain: the level of confidence is "low" to "medium."

Finally, natural ecosystems also will be affected, but precisely how is very poorly understood. The mix of plant and animal species in a given ecosystem will change, but in ways that are difficult to predict and with a lag of decades to centuries (high confidence). In addition, species that are currently endangered or vulnerable will become rarer or extinct (high confidence). The number of species affected depends on the degree of warming and regional changes in precipitation.

From Impacts to Economic Damages

A climate change policy could be designed to reduce or eliminate the impacts in the previous section, and the policy's economic benefit would be the value of the damages avoided. However, measuring those damages is even more uncertain than assessing climate change through GCMs and calculating the resulting environmental impacts.

The deepest and most intractable uncertainties arise because of the long atmospheric lifetimes of most greenhouse gases.[22] Global temperatures and hence the damage caused by climate change are determined by the concentration of greenhouse gases in the atmosphere—a stock of gases that is the result of the accumulation of emissions over many years. Eliminating one ton of carbon dioxide emissions today will lower global carbon dioxide concentrations for the next 150 years or so, albeit very slightly. The benefit of the reduction, therefore, is a long stream of small reductions in damages extending very far into the future. However, it is impossible to know for certain how people living a hundred years from now will feel about the impact of climate change. Their preferences may be different, and they almost certainly will have a different range of opportunities to adapt to climate change or to mitigate its effects.

With few exceptions, about the best that can be done is to assume that people in the future will have essentially the same preferences and options as people living today.[23] It is a reasonable assumption, but only because

---

22. See table 2-2.

23. One place where it is possible to take account of the difference between people living today and those living in a hundred years is in calculating the effect of income growth on the pattern of consumption. People in the future are likely to have substantially higher real incomes than people today, especially in developing countries, and there are abundant historical data to show how their consumption patterns are likely to change as a result. Expenditure on food, for example, increases less than in proportion to income, while expenditure on medical care increases more than in proportion to income.

no alternative is more compelling. At the same time, it is almost certainly wrong: how well would people in 1900 have been able to predict what people in 2000 would be willing to pay for improvements in environmental quality? Any valuations assigned to climate damages far in the future, therefore, must be regarded as highly uncertain.

A further complication arises because the benefits of a reduction in emissions are spread out over time. The cost of the reduction, on the other hand, will be incurred at a single point: when the reduction takes place. In order to balance the cost of the reduction against the damages that will be avoided, a present-value calculation must be done to convert the stream of avoided damages into an equivalent one-time payment at the time of abatement. However, there is considerable uncertainty over the appropriate interest rate to use for a calculation of this type.[24]

Even without intertemporal complications, it would be difficult to measure the damages caused by higher temperatures with precision. Attaching dollar values to some impacts is relatively straightforward: for example, calculating the value of a change in net energy demand. Assigning values to others is more difficult. The agricultural damage done by climate change, for example, depends on the costs of adapting crops and farming methods, which vary across regions and are largely unknown. Most difficult of all is assigning a value to changes that are not mediated by markets, such as the extinction of a species or a change in an ecosystem. Many people clearly wish to avoid such things and are willing to pay to prevent them—or are willing to pay simply to avoid making an irreversible change in the climate; the problem is to determine how much. Economists do not even agree on the methodology to use.[25]

Because of these difficulties virtually no large-scale, detailed studies have been done of the economic impacts of climate change. Highly aggregated

24. The standard pareto efficiency justification for present-value calculations suggests that the rate should be closely linked to the market rate of return on physical capital. However, some economists disagree and advocate other interest rates or even other approaches to discounting.

25. Contingent valuation is the main method used to determine what people are willing to pay for environmental goods that they do not use directly. However, there are serious problems with contingent valuation: see Diamond and Hausman (1994) for a detailed critique.

studies have been done by Nordhaus (1992, 1994) and Cline (1992), and individual impacts have been studied by a number of authors, but there are many, many significant gaps. The most comprehensive study to date is IPCC (2001b), which surveyed the literature and reached several conclusions that are most notable for their uncertainty. The IPCC's overall assessment was that the aggregate market-sector impact of a small increase in global temperatures could be "plus or minus a few percent of world GDP," estimated with "medium confidence."[26] To put that in context, the IPCC's estimate of world gross domestic product (GDP) in 2050 is 59 to 187 trillion 1990 U.S. dollars, so if "a few percent" might mean 3 percent, the global damages from climate change could be plus or minus $5.6 trillion, or about the entire GDP of the United States in 1990. What's more, the IPCC's confidence in that statement is only "medium," meaning that the chance that the damage will be outside that range is 33 to 67 percent. Other conclusions are similarly uncertain: the aggregate non-market impacts could be "negative" (low confidence); most developed countries would have small net impacts, and some would actually benefit from warming (medium confidence); developing countries would be more vulnerable to climate change than developed countries (high confidence) and likely to suffer more adverse impacts (medium confidence); and larger temperature increases would cause aggregate effects to become increasingly detrimental in all countries (medium confidence). These findings indicate how little is really known about what economic damages might result from climate change.

What Are the Costs of Policies to Limit Emissions?

The costs of policies that would limit greenhouse gas emissions also are highly uncertain. The reason, in large part, is that baseline emissions are very difficult to predict.[27] The cost of reducing emissions to a target level depends heavily on how much they would grow otherwise: the more

26. IPCC (2001b), *Technical Summary*, p. 70. See footnote 21 for an explanation of the meaning of "medium confidence."

27. Baseline emissions are the greenhouse gas emissions that would occur in the absence of a climate change policy.

quickly emissions grow in the absence of a policy, the larger will be the reductions needed to reduce them to a given target. Not only that, but when baseline emissions are growing rapidly, reductions will have to be made sooner and therefore will be more expensive in present-value terms.

Many factors affect the baseline path of the world economy: the rate of population growth in different countries; the age structure, educational attainment, and economic productivity of populations; the rate of productivity growth within individual industries; the rate of (or lack of) convergence of developing countries' income and productivity to the levels prevailing in developed countries; production decisions of the Organization of the Petroleum Exporting Countries (OPEC); new developments in the technology of fossil fuel extraction; technical progress in conservation and fuel efficiency;[28] the discovery of new fuel deposits and reserves; and even the degree of social and economic integration among countries. As a result, the economy is very difficult to predict over long spans of time, and past attempts have generally been very far off the mark.[29] Plausible alternative assumptions about these factors can lead to vastly different estimates of emissions trajectories.

Even if the baseline path of the economy could be predicted perfectly, however, there would still be important uncertainties in calculating the cost of reducing emissions. Many key economic parameters are not known precisely. The scope of the problem can be conveyed by listing a few of these parameters: the short- and long-term price elasticity of demand for different fuels; the rate at which the composition of household demand changes as incomes rise;[30] the degree of substitutability of products from different countries; the intertemporal elasticity of substitution (which plays an important role in determining savings and capital formation); and the elasticity of the supply of labor. In addition, there is evidence that some inexpensive, efficient energy technologies already

28. This has been referred to as the "autonomous rate of energy efficiency improvement" (AEEI) in the literature on the cost of reducing climate change. There is very wide disagreement about its magnitude.

29. A notorious example was *Limits to Growth* (Meadows and others, 1972), the predictions of which verged on apocalyptic.

30. That is, the degree of nonhomotheticity in consumption.

exist that currently are not used, although the reasons that they have not been are very poorly understood.[31]

A final factor that must be considered in computing the costs of a climate change policy is any offsetting indirect benefits the policy produces. Some may be environmental: a climate policy that reduces fossil fuel consumption may also lower conventional air pollutants, such as sulfur dioxide, oxides of nitrogen, carbon monoxide, particulates, and volatile organic compounds (precursors of ground-level ozone). In some urban areas, the benefits could be substantial. Other benefits could be fiscal: if the climate change policy were a tax, for example, it would raise government revenue, perhaps allowing other taxes to be reduced. Such reductions could benefit the economy by increasing the supply of labor or by stimulating investment and capital formation. This has become known as the "double dividend" hypothesis, which has generated a considerable literature. However, the magnitude of any such effect is widely disputed and highly uncertain.

In spite of these uncertainties, a variety of studies have been done, most focusing on the near-term costs—through 2010 or 2020—of one of two policies: reducing emissions to 1990 levels or implementing the 1997 Kyoto Protocol.[32] Marginal costs typically are measured by calculating the carbon tax (a tax on the carbon content of fossil fuels) needed to achieve a particular emissions target. The results vary substantially across models.

Table 2-7 shows the carbon taxes needed in four regions, the United States, Europe, Japan, and CANZ (Canada, Australia, and New Zealand) in order for each region to achieve its 2010 Kyoto emissions target without any international permit trading.

Figure 2-11 illustrates the range of these results. It shows the median tax for each region, with error bars indicating the 20th and 80th percentiles in the distribution of results. The models agree most closely on the U.S. and CANZ carbon taxes, but the range is still large: the gap between the 20th and 80th percentiles for the United States is $153, and the gap between the smallest and largest results overall is $334. The

31. See IPCC (2001c), chapter 5.
32. The Kyoto Protocol will be discussed in detail in chapters 3 and 4.

Table 2-7. *Carbon Taxes Needed in 2010 to Achieve*
Kyoto Targets[a]

Model	U.S.	Europe	Japan	CANZ
ABARE-GTEM	322	665	645	425
AIM	153	198	198	234
CETA	168			
G-Cubed	76	227	97	157
GRAPE		204	304	
MERGE3	264	218	500	250
MIT-EPPA	193	276	501	247
MS-MRT	236	179	402	213
Oxford	410	996	1074	
RICE	132	159	251	145
SGM	188	407	357	201
WorldScan	85	20	122	46
Administration	154			
EIA	251			
POLES	136	135	195	131
Mean	198	307	387	205
Standard deviation	92	270	273	100

Source: IPCC (2001c), which draws heavily on Energy Modeling Forum 16, a multi-model evaluation of the Kyoto Protocol. The results of the study appear in a special issue of the *Energy Journal* (1999).

a. 1990 U.S. dollars per ton of carbon.

results for Europe and Japan are much less certain: the gaps between the 20th and 80th percentiles for those regions are \$314 and \$379, respectively.

The effect of these taxes on each region's GDP in 2010 is summarized in figure 2-12, which shows the median GDP loss for each region, along with the 20th and 80th percentiles of the distributions. In all regions, GDP losses are substantial, but once again, the ranges of results are large.

Uncertainty is the Central Feature of Climate Change

In summary, uncertainty is the single most important attribute of climate change as a policy problem. From climatology to economics, the

Figure 2-11. *Median Carbon Tax Needed in 2010 to Achieve Kyoto Target, by Region*[a]

1990 U.S. dollars per ton

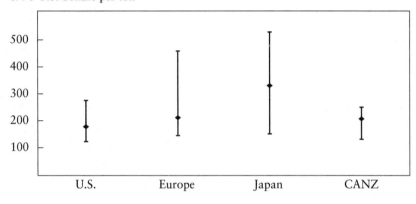

Source: Authors' calculations based on data from IPCC (2001c).
a. Error bars show the range between the 20th and 80th percentiles.

Figure 2-12. *Median GDP Loss in 2010 under Kyoto Targets, by Region*[a]

Percent of 2010 GDP

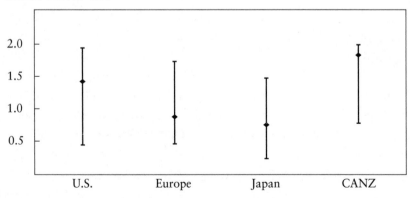

Source: Authors' calculations based on data from IPCC (2001c).
a. Error bars show the range between the 20th and 80th percentiles.

uncertainties in climate change are pervasive, large in magnitude, and very difficult to resolve. However, that does not mean that the appropriate policy is to do nothing. Instead, what is required is a coordinated international approach that reduces emissions of greenhouse gases when and where it is cost effective to do so.

3 The History of International Negotiations

International negotiations on climate change policy began in earnest in 1992 at the Rio Earth Summit, organized by the United Nations. The result was the United Nations Framework Convention on Climate Change (UNFCCC), a nonbinding agreement aimed at reducing atmospheric concentrations of greenhouse gases to achieve the goal of "preventing dangerous anthropogenic interference with the Earth's climate system."[1] It was signed and ratified by most of the countries in the world, including the United States, and went into force in 1994.

The convention's intent was to stabilize emissions of greenhouse gases at 1990 levels by the year 2000 through voluntary measures taken by individual countries. Most of the burden was to be assumed by forty industrialized countries listed in annex I to the convention. In particular, article 4, paragraph 2(a), required each of these countries to

---

1. For more information about the UNFCCC and the various COP meetings that followed it, see the UNFCCC website: www.unfccc.org.

"adopt national policies and take corresponding measures on the mitigation of climate change" in order to reduce its emissions. Annex I countries also were required to contribute to a financial fund, subsequently merged into the Global Environment Facility (GEF), to be used to help pay for climate-friendly projects in developing countries.

In the subsequent decade, however, few substantive policies were implemented and global emissions of greenhouse gases rose considerably. From that perspective, the UNFCCC failed to achieve its goal. However, its real contribution was to establish a mechanism through which negotiations would continue at periodic "conference of the parties" (COP) meetings.

The Berlin Mandate

The first conference of the parties (COP 1) was held in Berlin in March and April of 1995. The main concern of participants was that the UNFCCC seemed to be having little effect on greenhouse gas emissions. The problem was thought to be article 4, paragraph 2(a), which required annex I countries to "adopt national policies and take corresponding measures on the mitigation of climate change," a vague directive that imposed no specific policies or targets and required little more than good intentions from participants. The goal of the meeting was to begin a new round of negotiations that would lead to a stronger international protocol or other legal instrument. To that end, the conference produced an agreement, the Berlin Mandate, that established a two-year process of review to strengthen the commitments made by developed countries in the UNFCCC. The review was intended to identify and describe specific "policies and measures" that could be taken by developed countries and to "set quantified limitation and reduction objectives within specified time-frames" for those countries. No new commitments or obligations were imposed on developing countries.

The second conference, COP 2, was held in Geneva in July 1996. It reviewed and endorsed the findings in the IPCC's *Second Assessment Report* (IPCC 1995) and surveyed progress on the Berlin Mandate. The meeting ended with a declaration that rephrased the second task of the mandate in a small but significant way: the goal was now to determine "quantified *legally-binding* objectives for emission limitations and significant overall reductions within specified time-frames" for developed countries (emphasis added). This laid the groundwork for COP 3, which was intended to produce a much stronger climate change agreement in which developed countries agreed to specific, binding emissions targets and timetables for achieving them.

The Kyoto Protocol

COP 3 was held in Kyoto in December 1997. The result was the Kyoto Protocol, a treaty that formalized the targets and timetables approach that had been taking shape since COP 1. The protocol set explicit emissions targets for the thirty-nine countries listed in its annex B, which included essentially all industrialized countries that were signatories.[2] Each of these countries was to reduce its greenhouse gas emissions so that its total emissions, when converted to a carbon-equivalent basis, did not exceed a specified percentage of its base period emissions. For most countries the base period was 1990, but countries having economies in transition were allowed to choose another base period during COP 2.[3] Average emissions over the budget period 2008–12 were to be at or below the target.[4] The annex B limits are shown in table 3-1.

The commitments in table 3-1 amount to about a 5 percent reduction below 1990 emissions for the annex B countries as a group, or about 245 million metric tons of carbon.[5] As shown in table 3-2, however, they are *not* a reduction below *current* annex B emissions. Several political events have reduced annex B emissions substantially since 1990 for reasons unrelated to climate change. Major restructuring of the coal industry in the

2. The annex B list is a subset of the countries listed in annex I of the UNFCCC. It excludes Belarus, which had not ratified the UNFCCC by the time COP 3 was held, and Turkey, which requested that it be removed from annex I at COP 3.

3. Decision 9 of COP 2 established the base periods for annex I countries.

4. Gases other than carbon dioxide are converted to a carbon-equivalent basis using global warming potentials such as those shown in table 2-2. A country's carbon-equivalent emissions over the five-year period 2008–12 were required to be less than or equal to the specified fraction of its base-period emissions.

5. The exact reduction depends on the treatment of land use changes, which had not been finalized by the end of COP 6.

Table 3-1. *Kyoto Protocol Emissions Limits*
or Reduction Commitments[a]

Country	Target	Country	Target
Australia	108	Liechtenstein	92
Austria	92	Lithuania[b]	92
Belgium	92	Luxembourg	92
Bulgaria[b]	92	Monaco	92
Canada	94	Netherlands	92
Croatia[b]	95	New Zealand	100
Czech Republic[b]	92	Norway	101
Denmark	92	Poland[b]	94
Estonia[b]	92	Portugal	92
European Community	92	Romania[b]	92
Finland	92	Russian Federation[b]	100
France	92	Slovakia[b]	92
Germany	92	Slovenia[b]	92
Greece	92	Spain	92
Hungary[b]	94	Sweden	92
Iceland	110	Switzerland	92
Ireland	92	Ukraine[b]	100
Italy	92	United Kingdom	92
Japan	94	United States	93
Latvia[b]	92		

a. Percent of 1990 or base period emissions.
b. Country designated an "economy in transition."

United Kingdom and of much of the German economy after unification have reduced emissions in both countries substantially below their Kyoto commitments. More important, however, the collapse of the Russian economy has reduced Russian carbon emissions by nearly 300 million metric tons compared with emissions in 1990, and similar changes have occurred in other parts of the former Soviet Union.[6] As a result, annex B emissions in 1998 were about 6 percent below 1990 levels, or about 70 million metric tons of carbon below the Kyoto target.[7] Emissions could continue to grow for several years before the annex B constraint became binding.

6. The gap between Russia's commitment and its current emissions is often referred to as "hot air" because it could be sold to other countries without any emissions reductions in Russia.
7. The exact values depend on the treatment of land use changes.

Table 3-2. *Kyoto Greenhouse Gas Emissions in 1990 and 1998*[a]

Country	1990	1998	Change[b]
Australia	115	132	17
Austria	21	22	1
Belgium	37	40	2
Bulgaria	43	23	–20
Canada	167	189	22
Czech Republic	52	40	–12
Denmark	19	21	2
Estonia	11	6	–5
Finland	21	21	0
France	151	152	1
Germany	330	278	–52
Greece	29	34	5
Hungary	28	23	–5
Iceland[c]	1	1	0
Ireland	15	17	3
Italy	141	148	6
Japan[c]	331	363	32
Latvia	10	3	–7
Liechtenstein[c]	0	0	0
Lithuania	14	7	–7
Luxembourg[c]	4	3	–1
Monaco	0	0	0
Netherlands	59	64	5
New Zealand	20	20	0
Norway	14	15	1
Poland	154	110	–44
Portugal	17	20	3
Romania[c]	72	45	–28
Russian Federation[c]	829	535	–294
Slovakia	21	14	–6
Slovenia[c]	5	5	0
Spain	83	101	17
Sweden	19	20	1
Switzerland	14	15	0
Ukraine	251	124	–127
United Kingdom	202	185	–17
United States	1,650	1,835	185
Total	4,949	4,631	–318

Source: Table B.1 in UNFCCC emissions data file "ghgtabl90-98.zip" dated November 1, 2000. Figures include emissions of all gases covered by the protocol but do not include the effects of land use changes or forestry.

a. Millions of metric tons of carbon.

b. Values may not equal the difference between 1990 and 1998 values because of rounding error.

c. Data for 1998 are not available. Emissions are reported for the latest available year instead: Iceland 1995; Japan 1997; Liechtenstein 1990; Luxembourg 1995; Romania 1994; Russian Federation 1996; and Slovenia 1990.

The protocol was designed to allow annex B countries flexibility in meeting their commitments; some of the flexibility concerns the unilateral actions countries can take to comply with the protocol. First, the specific policies to be used to reduce emissions are left completely to the discretion of each country. Second, compliance can be achieved by any mix of carbon-equivalent reductions in four individual gases and two classes of halocarbon: carbon dioxide, methane, nitrous oxide, sulfur hexafluoride, hydrofluorocarbons (HFCs), and perfluorocarbons (PFCs). Third, countries can offset some of their emissions by enhancing "sinks" of carbon dioxide: forests or other mechanisms that remove carbon dioxide from the atmosphere. Fourth, reductions that exceed annex B commitments can be carried forward and used to count toward compliance in future periods.

The protocol also provides three mechanisms that allow for flexibility on a multilateral basis. The most important is international emissions permit trading (IET), which is allowed among annex B countries under the protocol's article 17. In addition, article 6 of the protocol allows for joint implementation (JI), a project-based system under which one annex B country can receive credit for emissions-reducing activities it finances in another annex B country. The use of emissions trading and JI, however, must be "supplemental to domestic actions," a vague phrase that leaves open the possibility of imposing quantitative limits on the amount of trading and joint implementation.[8]

Finally, article 12 of the protocol created the clean development mechanism (CDM), through which an annex B country can receive credit for emissions-reducing projects it undertakes in a developing (non–annex B) country. The project must produce "certified emissions reductions" that result in "real, measurable, and long-term benefits related to mitigation of climate change." The reductions would have to be certified by a designated international agency as "additional" to any reductions that would have occurred in the absence of the project.

For the protocol to come into force it must be ratified by 55 percent of its signatories, and they must jointly account for at least 55 percent of

---

8. The European Union, in particular, was in favor of limiting the degree to which compliance could be achieved by trading and joint implementation. The United States was opposed to any restrictions.

total carbon dioxide emissions in 1990 from annex I countries. Most of the operational details of all three international flexibility mechanisms were left for future COP meetings to resolve. There was no negotiation on issues of compliance, on how institutional structures would work, or on how developing countries might be involved apart from participating in a CDM project.

Meetings after COP 3 were devoted to working out the operational details of the Kyoto Protocol. COP 4, held in Buenos Aires in November 1998, adopted a two-year work program intended to produce a detailed set of recommendations on the operation of emissions trading, joint implementation, and the clean development mechanism that could be adopted at COP 6. Within the program, priority was given to an issue important in securing the participation of developing countries: the funding mechanism incorporated into the Global Environment Facility and the CDM. The operation of the CDM was designed "with the objective of ensuring transparency, efficiency and accountability through independent auditing and verification of project activities." COP 4 also clarified, to an extent, the accounting rules to be used for sinks.

COP 5, held in Bonn in 1999, was devoted mostly to monitoring the progress of the work program adopted at COP 4. The institutional design of the JI and CDM mechanisms were discussed, along with measures that could be adopted to penalize noncompliance. However, decisions on these matters were postponed to COP 6.

The Bonn and Marrakesh Agreements

The goal of COP 6, begun at the Hague in November 2000 and concluded at Bonn in July 2001, was to resolve all remaining implementation details of the Kyoto Protocol. The specific rules of emissions trading, joint implementation, and the clean development mechanism were to be determined, along with further details on the role of sinks. In addition, the meeting was to adopt a mechanism that would ensure compliance once the protocol came into force.

Deep disagreements among the participants could not be resolved during the November meeting. Some of the key issues in dispute were the size and role of sinks; the interpretation of the "supplementarity" restrictions

Table 3-3. *Countries Receiving Sink Allowances Exceeding 1 Million Metric Tons*[a]

Country	Allowance[a]
Canada	12.00
Germany	1.24
Japan	13.00
Romania	1.10
Russia	17.63
Ukraine	1.11

a. Million metric tons of carbon.

on trading and joint implementation; and the design of compliance mechanisms. The parties decided to adjourn and reconvene in a second session of COP 6 to be held in Bonn in July 2001. Virtually all important decisions were postponed.

In March 2001, however, President George W. Bush announced that the United States would withdraw from the Kyoto Protocol. When COP 6 resumed in July, the delegates decided to proceed without the United States and assembled a package of decisions called the Bonn Agreements on the Implementation of the Buenos Aires Plan of Action. Some of the features of the Bonn Agreements were favored by the United States, including a decision not to subject the international flexibility mechanisms to quantitative restrictions.

In addition, the overall stringency of the protocol was reduced by granting countries forestry and land use sink allowances totaling 54.5 million metric tons (mmt) of carbon. Countries given sink allowances greater than 1 million metric tons of carbon-equivalent emissions are shown in table 3-3.

These allowances, combined with the fact that Russian emissions already are far below Russia's Kyoto allowance, are likely to mean that the protocol will not be binding in the 2008–12 budget period if the United States does not participate. A rough calculation demonstrates this point. Excluding the United States, the 1998 annex B greenhouse gas emissions shown in table 3-2 total about 350 mmt less than non-U.S. annex B commitments. Adding in the sink allowance means that non-U.S. emissions could rise by about 400 mmt before reaching the Kyoto constraint. Over the last decade, emissions of countries in the Organisation for Economic Co-operation and Development (OECD) have been growing at an average rate of about 1.2 percent per year. If that rate remains constant and emissions from non-OECD annex B countries grow at the same rate, the Kyoto commitments will not be binding until 2009. If the growth rate falls to 1 percent per year, the commitment level will not be reached until 2011. In contrast, *with* U.S. participation, the annex B constraint would be reached in less than three years.[9]

Although the Bonn Agreements were formulated at COP 6, they were not adopted as official decisions of the conference; instead, further discussion and formal adoption were deferred until COP 7, which was held in Marrakesh in October and November 2001. It refined and extended the Bonn Agreements in three main areas: defining the "principles, nature and scope" of the international flexibility mechanisms; finalizing the accounting rules for sinks derived from land use changes and forestry; and designing an enforcement mechanism to discourage noncompliance. The result was the Marrakesh Accords, which COP participants hoped would remove all remaining obstacles to ratification of the Kyoto Protocol.

The accords include detailed specifications for many of the institutional and administrative aspects of the international flexibility mechanisms. It defines four classes of emissions permits: AAUs, ERUs, CERs, and RMUs. One unit of each of these allows one ton of carbon-equivalent emissions. The main difference between them is their institutional origin. AAUs (assigned amount units) originate with annex B commitments: each country is entitled to issue an amount of AAUs equal to its commitment. ERUs (emissions reduction units) are generated by joint implementation projects. A JI host country (where the emissions-reducing project takes place) converts an appropriate number of its AAUs to ERUs on a one-to-one basis and then transfers them to the country sponsoring the project. CERs (certified emission reductions) are generated by CDM projects. RMUs (removal units) are generated by sink enhancements; each country's sink allowance specifies the maximum number of RMUs it may generate. Unused AAUs may be carried forward from one budget period to

9. This calculation assumes that the United States would be granted 28 mmt of sink allowances, which is the amount it had requested before withdrawing from COP 6.

the next without restriction. Unused ERUs and CERs may be carried forward but only up to a maximum of 2.5 percent of their owner's AAUs. Unused RMUs may not be carried forward at all.

The design and operation of the clean development mechanism are also spelled out in detail, as are the accounting rules to be used for tracking emissions permits. In places, the level of detail is remarkable: there are, for example, exact specifications regarding the way serial numbers are to be assigned to emissions permits. In other areas, however, the accords remain vague. The most important of these areas is the compliance mechanism to be used to enforce the protocol. The fundamental sanction that would be imposed on a country exceeding its emissions allowances during the 2008–12 budget period would be to reduce its AAU allotment in the next budget period by 1.3 times its violation. However, base allowances for the post-2012 period have not been negotiated, so any country exceeding its commitment in 2008–12 could simply demand a higher base allowance in the next period. Moreover, enforcement is to be handled by an "enforcement branch," which has yet to be designed.

Although the Marrakesh Accords essentially eliminate the supplementarity restrictions that had been proposed for the Kyoto trading mechanisms, they do limit trading in one important respect: annex I countries must keep "in reserve" a number of permits equal to at least 90 percent of their annex B commitment or five times their most recent year's emissions, whichever is lower. (The factor of five arises because the budget period is five years long.) The intent of this provision is to prevent countries from selling large numbers of permits and then failing to curb their emissions by a corresponding amount. The difference between this and supplementarity is that supplementarity would limit the number of permits a country could buy while this restriction would limit the number it could sell.

Finally, COP 7 further relaxed the Kyoto emissions target by granting a Russian request that its sink allowance be increased from 17.63 mmt to 33 mmt. As a result, it is even less likely that the treaty will constrain emissions during the 2008–12 budget period in the absence of U.S. participation.

4 | Why the Kyoto Protocol Is the Wrong Approach

Several COP meetings and years of negotiations have been devoted to the Kyoto Protocol, but it remains unlikely ever to succeed in reducing greenhouse gas emissions. At most, it will end up a paper tiger: an agreement that looks strong on the surface but has no viable mechanism for enforcement and does little or nothing to control emissions. It will never be more than that because the fundamental principle on which it is based—setting "targets and timetables" for reductions in greenhouse gas emissions—is economically flawed and politically unrealistic. The core problem is straightforward: in order to ratify the protocol, a developed country must agree to reduce its emissions to a specified level by a particular date, regardless of the cost. Since the cost could be very large, few countries with substantial greenhouse gas emissions will ratify the treaty and those that do will be unlikely to comply with its requirements. Developing nations, which will become the world's largest emitters in coming decades, will have even less incentive to adopt commitments. Long-term participation in the protocol will end up being limited to countries that account for only a small share of world greenhouse gas emissions.

The theoretical flaws in the targets and timetables approach will be examined in detail in chapter 5, which discusses how uncertainty should be incorporated into the economic design of a sound climate policy. The practical problems, however, can be seen very easily by considering some of the key issues that would be raised by prudent, unbiased legislators if the protocol is presented for ratification.

Costs, Benefits, and Uncertainty

The first issue would be whether the protocol passes a cost-benefit test. This already has been an important part of the debate in a number of countries, although in a somewhat disguised form. Proponents of the treaty tend to argue that the benefits must be greater than the costs because the damages from climate change could be so severe. Opponents argue that the costs of the policy are likely to be large and its benefits uncertain and possibly very small.[1]

Neither position is correct, although each has elements of truth. In light of the vast uncertainties discussed in chapter 2, an honest assessment must acknowledge that it is impossible to say for certain whether the benefits of the treaty would outweigh its costs. Studies to date provide little justification for the protocol. It would only reduce the rate of warming slightly, not prevent it entirely. As a result, its benefits would be small compared with the estimated damages from uncontrolled warming. In addition, the protocol's effect on temperatures would be decades away while its costs would begin immediately. Thus, for most developed countries, including the United States, the protocol provides only small environmental benefits but imposes significant costs. For example, Nordhaus and Boyer (1999) find that the protocol does not "bear any relation to an economically oriented strategy that would balance the costs and benefits of greenhouse gas reductions." They calculate that the worldwide present value cost of the Kyoto Protocol would be $800 to $1,500 billion if it is

---

1. The extreme form of this view is the assertion by some political groups that global warming is an "unproven theory."

implemented as efficiently as possible while the present value of benefits would be $120 billion. Other studies reach similar conclusions. Tol (1999), for example, finds that the Kyoto Protocol would have a net present value cost in excess of $2.5 trillion and comments that "the emissions targets agreed in the Kyoto Protocol are irreconcilable with economic rationality."

The central flaw in the protocol is its emphasis on targets and timetables for emissions reductions. Because each country must meet its emissions commitment regardless of the cost, there is no way for a government to be sure that adopting the protocol would not be disastrously expensive. As a result, governments of the countries that matter most in climate change—those with substantial greenhouse gas emissions—will be very reluctant to commit themselves to the protocol. To do so would be an open-ended surrender of a potentially important element of their sovereignty.

Proponents of the protocol downplay this issue by arguing that the protocol's international flexibility mechanisms—international emissions trading, joint implementation, and the clean development mechanism—will keep abatement costs low. However, this misrepresents the effect of the mechanisms: although flexibility reduces overall abatement costs, it does *not* guarantee that costs will end up being acceptably small. The real determinant of overall costs is the degree of emissions reduction achieved by the policy: the tighter the constraint, the higher the costs. If the emissions constraint is sufficiently tight, costs will be high even with the international flexibility mechanisms in place.

For example, if OECD countries were to stabilize their carbon emissions at 1990 levels (a target that is less stringent than the protocol's),[2] international permit trading would reduce overall costs by 13 to 15 percent relative to the case in which each country had to stabilize its emissions on its own.[3] There clearly are gains from trade, but they are a relatively

2. The list of OECD countries differs somewhat from the list of industrialized countries subject to obligations under the Kyoto Protocol. Four OECD countries do not have obligations under Kyoto: Korea, Mexico, the Slovak Republic, and Turkey. Eleven non-OECD members do have obligations: Monaco, Lichtenstein, and nine countries from eastern Europe and the former Soviet Union.

3. McKibbin, Shackleton, and Wilcoxen (1999).

small share of overall costs. Studies of the protocol tend to find much larger gains, but most of the improvement is due to the fact that annex B trading relaxes the overall emissions constraint substantially. When trading is allowed, Russia is able to sell its 300 million metric tons of excess emissions permits to other annex B countries, and annex B countries as a group do that much less abatement. Abatement costs fall simply because less abatement is being done.

International Transfers of Wealth

Besides failing to guarantee that costs will be low, the international flexibility mechanisms in the protocol have a potentially serious political liability: they can lead to large transfers of wealth between countries.[4] Moreover, as shown in appendix A, these transfers are likely to exceed the efficiency gains from trading by a factor of 10 or more. A net cost reduction of $1 billion, for example, could easily require a $10 billion transfer.[5]

To illustrate the magnitude of the potential transfers, consider the following rough calculation. In 1990 the United States emitted about 1,340 million metric tons of carbon in the form of carbon dioxide. Carbon emissions have been growing over time, so suppose that by 2010 the United States ended up needing to import permits equal to about 20 percent of 1990 emissions, about 268 million tons. There is enormous uncertainty about what the price of an international carbon permit might be, but $100 per ton is well within the range of estimates; some studies have projected prices of $200 or more. Under these conditions, the permit system would add $27 billion to $54 billion to the U.S. trade deficit every year. The transfer would become even larger over time as the price of emissions permits rose. To put these figures in context, the entire U.S. trade deficit in 1996 was $114 billion, and the value of permits would dwarf the U.S. foreign aid budget, which is now about $15.3 billion.

---

4. The more important permit trading is for keeping costs low, the larger are the international payments it will generate. Appendix A examines this point in detail.

5. For example, a country might avoid $11 billion in abatement costs by buying permits worth $10 billion. There would be a $1 billion gain from trade and a $10 billion transfer.

Moreover, permit exchanges of this magnitude would put enormous stress on the world trade system. The balance of trade for a country importing permits would deteriorate substantially. Although trading would equalize marginal abatement costs across countries, achieving that efficiency would be likely to lead to substantial volatility in exchange rates and distortions in the world trade system.

Enforcement

A related difficulty with the protocol, acknowledged even by its supporters, is that no individual government would have any incentive to police it. It is easy to see why: monitoring polluters is expensive and punishing violators imposes costs on domestic residents in exchange for benefits that accrue largely to foreigners. There would be a strong temptation for governments to look the other way if firms exceeded their emissions permits. For the treaty to be effective, therefore, it will need to include a strong international mechanism for monitoring compliance and penalizing violations. Devising such a mechanism has been a key obstacle in the ongoing COP negotiations.

Moreover, the protocol's international permit trading system makes monitoring and enforcement especially important. Ordinarily, the effect of cheating on an international agreement is simply to undermine the agreement's goal: in this case, by allowing emissions to be higher than intended in the treaty. Under the protocol, however, cheating has a second effect: it debases the value of emissions permits and fragments the international permit market.

In an ideal world, permits from different countries would be perfectly homogeneous goods: a permit from country A would be identical to a permit from country B. Buyers would not prefer one to the other, and both would trade at a single price because no one would be willing to pay more for a permit from country A than for one from country B and vice versa. In this situation, the permit market would ensure that emissions were reduced at the minimum total cost. Firms that could reduce their emissions at low cost would find it profitable to do so in order to be able to sell permits. Firms facing especially high cleanup costs, on the other hand, would find it more profitable to buy the permits and do less abatement. In the end, the abatement would be done by firms that could do it at the lowest cost.

When cheating is possible, however, permits sold by different countries could become heterogeneous assets that no longer trade at identical prices. To see why, suppose that a country fails to enforce the protocol within its borders and its firms sell some or all of their emissions permits on the international market without making corresponding reductions in their emissions. If the violation is later detected by the international monitoring agency, those permits could be declared invalid.[6] A firm that bought one of the permits would suffer a substantial windfall loss and might also find itself abruptly thrown out of compliance if its remaining permits were insufficient to cover its emissions. Permit buyers would be aware of this risk and would pay less for permits from countries perceived as being more likely to cheat. In essence, the potential for cheating means that each permit will carry a risk of default, the magnitude of which will depend on enforcement of the protocol in the country of origin.

When permit prices differ across countries, the permit trading system no longer ensures that emissions will be cleaned up at the minimum possible cost. For example, suppose that country B is thought to have a higher risk of cheating than country A. The difference in risk will mean that the equilibrium price of B's permits, P_B, will be lower than the price of A's permits, P_A. The permit prices will determine how much abatement is done in each country: firms in A will have an incentive to reduce emissions until the marginal cost of doing so rises to P_A, while firms in B will reduce emissions until marginal costs rise to P_B. Since P_A is greater than P_B, however, the last units abated in A will cost more than they would have if they had been abated in B instead. The total cost of abatement will be too high: costs would be lower if some of the abatement was shifted from A to B.

---

6. This is known in the climate policy literature as "buyer liability" because permit buyers lose when cheating is discovered. An alternative would be "seller liability," under which buyers would be able to keep and use permits bought in good faith and the seller of the permits would be penalized. The Marrakesh Accords adopt trading restrictions that essentially impose a form of seller liability on annex I countries; the rules on CDM transactions are less clear.

In sum, the protocol requires an especially strong international mechanism for monitoring and enforcement. Without it, the agreement will fail to achieve its abatement goals and the international permit market will be fragmented and inefficient. Implementing such a mechanism will be very difficult, and no viable proposals for doing so have been made to date.

Developing Country Participation

Another issue important to ratification that already has been raised in the debate is the role of developing countries. To have a significant long-term effect on greenhouse gas emissions, any agreement eventually must include substantial participation by developing countries. The UNFCCC and the Kyoto Protocol, however, are essentially silent on how that participation will come about.[7] The UNFCCC explicitly assigns all responsibility for near-term abatement to developed countries on the grounds that they have been the largest sources of greenhouse gas emissions to date.[8] The Kyoto Protocol goes one step further, requiring developed countries to adopt only climate policies that "minimize adverse effects . . . on other Parties, especially developing country Parties."[9]

Developing countries have been vigorously opposed to making commitments to reduce their own emissions. Not only have they been responsible for a relatively small share of historical emissions, they face even larger uncertainties about the future than developed countries do. As a result, they are even more reluctant to give up any of their sovereignty in a climate change agreement.

---

7. The Kyoto Protocol does include developing country participation through the clean development mechanism, but the CDM is designed to help *developed* countries meet their commitments. As yet it places little effective restriction on overall emissions of participating developing countries.

8. Article 3, paragraph 1, of the UNFCCC states, in part, that "the developed country Parties should take the lead in combating climate change and the adverse effects thereof." The rationale is contained in the UNFCCC's preamble, which states, in part, that "the largest share of historical and current global emissions of greenhouse gases has originated in developed countries [and] per capita emissions in developing countries are still relatively low and that the share of global emissions originating in developing countries will grow to meet their social and development needs."

9. Kyoto Protocol, article 2, paragraph 3.

Proponents of the protocol argue that in time, developing countries will voluntarily accept responsibility for limiting their greenhouse gas emissions as their per capita income rises. However, the only incentive for a developing country to join annex I and undertake a specific emissions reduction commitment would be to participate in the international emissions permit trading system. If there are large differences in abatement costs between developed and developing countries and if developing countries have sufficiently large greenhouse gas allowances, it would be possible for them to earn foreign currency by becoming exporters of emissions permits. Essentially, developed countries would pay developing countries for abatement.

Joining annex I would present risks, however. Massive exports of permits would lead to appreciation of the country's exchange rate and a decline or collapse in its exports other than permits. Also, the permit revenue comes with strings attached: much of it would have to be invested in improved energy technology in order to reduce emissions and free up the permits in the first place. This is unlikely to be an ideal strategy for long-term economic development and would make the policy unattractive to developing countries.

Prospects for the Protocol

Ironically, even if the Kyoto Protocol were ratified immediately, it may not constrain emissions for years to come. Emissions from the United Kingdom, Germany, and especially Russia are below 1990 levels already. The reasons are varied, but they have nothing to do with climate change policy. Emissions in the United Kingdom dropped as a result of changes in its coal industry begun under the Thatcher government; German emissions fell because reunification led quickly to the elimination of many energy-inefficient activities in what was East Germany; and Russian emissions were reduced because the Russian economy collapsed in the 1990s. As a result, total emissions from annex B countries currently are below 1990 levels. If the protocol goes forward without the United States, emissions from the remaining countries would be about 400 million metric tons below the target. It is unlikely that emissions would be significantly constrained during the protocol's first commitment period, 2008 to 2012.

Moreover, the protocol's emissions targets apply *only* to the 2008–12 period: limits for future periods remain to be negotiated. If it fails to constrain emissions in the first commitment period, the protocol will have done nothing to reduce the risks posed by climate change.

In sum, the Kyoto Protocol is badly flawed, and it will do little to reduce greenhouse gas emissions. Indeed, its flaws are precisely the reason that years of negotiations have gone by while greenhouse gas emissions have continued to grow unchecked. It fails to address the deep uncertainties in climate change; it is designed around international permit trading, a politically unrealistic mechanism given the transfers of wealth that it could cause; it has no viable mechanism for enforcement; and it contains no realistic process for increasing participation by developing countries. Countries that are major sources of greenhouse gases are unlikely to ratify the agreement unless their own commitments are very lax, or they will fail to comply with it when their constraints become binding. It is an impractical policy focused on achieving an unrealistic and inappropriate goal.

5 Designing a Realistic Alternative

The uncertainties associated with climate change have polarized public debate. Some observers argue that the uncertainties are too large to justify immediate action—that climate change is an "unproved theory"—and that the best response is to do more climate research and wait for the uncertainties to be resolved. Other observers take the opposite position, maintaining that the risks from global warming are so severe that substantial cuts must be made in greenhouse gas emissions immediately, regardless of the cost. Neither position is appropriate. On one hand, increasing the concentration of greenhouse gases in the atmosphere exposes the world to the risk of an adverse change in the climate, even though the distribution of that risk is poorly understood. Enough is known to justify reducing greenhouse gas emissions, particularly to preserve the option of avoiding an irreversible change in the climate. On the other hand, too little is known about the causes and consequences of climate change to justify a draconian cut in emissions. Given the uncertainties, a prudent approach would be to abate emissions where possible to do so at modest cost.

Permits and Taxes under Uncertainty

Economic theory provides guidance regarding the most appropriate structure for a climate change policy. Because greenhouse gases are emitted by a vast number of highly heterogeneous sources, minimizing the cost of abating a given amount of emissions requires that all sources clean up an amount sufficient to equalize their marginal costs of abatement. To achieve this, the standard economic policy prescription would be to adopt a market-based instrument such as a tax on emissions or a tradable permit system for emission rights.

In the absence of uncertainty, the efficient level of abatement could be achieved under either policy, although the tax and emissions trading policies would have sharply different effects on different groups within the economy (distributional effects). Figure 5-1 illustrates this point. The horizontal axis shows the amount of abatement as a percent of uncontrolled emissions; when abatement reaches 100, emissions have been driven to zero. Two equivalent policies will move the economy to the efficient level of abatement: a tax on emissions at rate T or a permit policy with Q_p permits.

If tax T were placed on emissions, every source would clean up all emissions that could be abated at a marginal cost less than or equal to T, pushing the level of abatement to Q_a. If a permit policy were imposed, on the other hand, sources would be forced to clean up to Q_a because only Q_p emissions would be allowed. The policies are quite different in some respects, discussed further below, but in either case the efficient level of abatement, Q_a, would be achieved.

Under uncertainty, however, the situation becomes more complicated. In a path-breaking 1974 paper, Martin Weitzman showed that taxes and permits are *not* equally efficient when marginal benefits and costs are uncertain and that the relative slopes of the two curves determine which policy will be better. To see why, consider a hypothetical air pollutant. The pollutant is dangerous only at high levels: it causes no damage at all when daily emissions are below 100 tons, but each ton emitted beyond that amount causes $10's worth of health problems. Emissions currently are 150 tons per day, so the marginal benefit of abatement would be $10 (the damage avoided) for each of the first fifty tons eliminated. Beyond

Figure 5-1. *Emission Taxes and Permits*

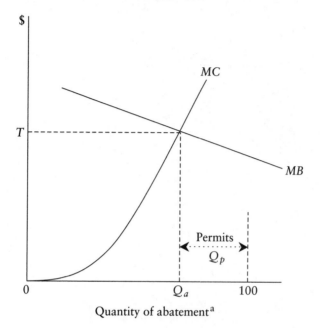

Quantity of abatement[a]

a. Percent of uncontrolled emissions.

that point, however, the marginal benefit of abatement would drop to zero: emissions would be below the 100-ton threshold and no longer cause any damage. This is an example of a steep marginal benefit curve: at the threshold, marginal benefits go rapidly from $10 to zero. Finally, suppose that the pollutant can be cleaned up with constant returns to scale—the marginal cost curve is flat—but that the precise cost is uncertain: all that is known is that the cost of cleanup is less than $10 per ton.

Given this information, the efficient amount of pollution is 100 tons. Above 100 tons, the damage of an additional ton is higher than the cost of abating it; below 100 tons, further reductions produce no additional benefit. In this situation, a permit policy would be far better than an emissions tax. By issuing permits for 100 tons of emissions, the government could be sure of achieving the efficient outcome: for any marginal cost below $10, the permit system will keep emissions from exceeding the

Figure 5-2. *Permit Policy in the Absence of Uncertainty*

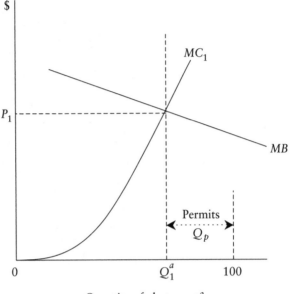

Quantity of abatement[a]

a. Percent of uncontrolled emissions.

threshold. A tax, on the other hand, would be a terrible policy. Suppose that the government imposed a $5 tax on the pollutant and that the marginal cost of abatement turned out to be $6. In that case, firms would choose to pay the tax rather than do any abatement and emissions would remain at 150 tons. If the cost turned out to be low, say $4, the situation would be no better: in that case, firms would clean up everything and emissions would drop to zero. This example captures the essence of the advantage permits have over taxes when marginal benefits are steep and marginal costs are flat: in that situation, it is important to get the quantity of emissions down to a threshold. A permit policy does exactly that.

To illustrate Weitzman's point graphically, suppose that marginal benefits are known and that marginal abatement costs are believed, *ex ante*, to be given by marginal abatement cost curve MC_1 in figure 5-2 and that a permit policy is put in place with Q_p permits issued.

Figure 5-3. *Permit Policy with Unexpectedly High Costs*

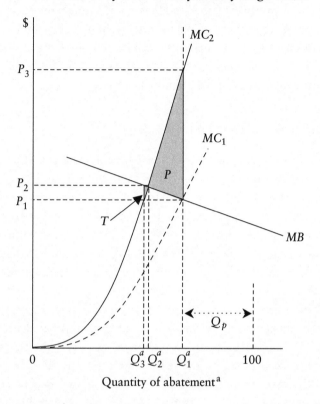

a. Percent of uncontrolled emissions.

As long as MC_1 is correct, the policy will be efficient and the level of abatement will be Q_1^a. The price of a permit will rise to P_1, the marginal cost of the last unit abated. However, suppose that after the policy is implemented, marginal costs are discovered to be much higher than expected. This situation is shown in figure 5-3, where MC_2 is the *ex post* marginal cost curve. Under these circumstances, the efficient level of abatement would fall to Q_2^a. Because there are only Q_p permits, however, firms will be forced to abate to Q_1^a. The price of a permit would rise substantially, to P_3, and the costs of the additional abatement would exceed the benefits by shaded triangle P in the diagram.

The potential inefficiency of a permit system is not just a theoretical curiosity: it is intuitively understood by many participants in the climate change debate, even those who have little training in economics. A non-economist might sum up a permit system by describing it as a policy that "caps emissions regardless of cost." The language differs from that an economist would use, but the point is the same.

Inframarginal Transfers under a Tax

In this example, a tax policy would have been a much wiser choice. Suppose that in the initial situation, with the low expected marginal cost curve, a tax had been imposed equal to P_1. Once firms discovered that the true marginal cost curve was substantially higher than expected, the level of abatement would drop to Q_3^a—slightly too low but with a much smaller welfare loss (triangle T) than with the permit policy. The advantage of a tax stems from the fact that the marginal cost curve is much steeper than the marginal benefit curve.

Applying this analysis to climate change policy suggests that a tax is likely to be far more efficient than a permit system. All evidence to date suggests that the marginal cost curve for reducing greenhouse gas emissions is very steep, at least for developed countries. At the same time, the nature of climate change indicates that the marginal benefit curve for reducing emissions will be very flat. The damages from climate change are caused by the overall stock of greenhouse gases in the atmosphere, which is the total emissions accumulated over many years. Greenhouse gases remain in the atmosphere for a long time: up to 200 years for carbon dioxide, 114 years for nitrous oxide, 45 to 260 years for chlorofluorocarbons, and up to 50,000 years for perfluoromethane (CF_4).[1] As a result, the marginal damage curve for emissions of a gas in any given year will be flat: the first ton and the last ton emitted in that year will have very similar effects on the atmospheric concentration of the gas and hence will cause very similar damages.[2] For example, any single year's emissions

1. The atmospheric lifetimes of common greenhouse gases are shown in table 2-2.
2. For more discussion of the benefits of abating emissions of stock pollutants, see Newell and Pizer (1998).

of carbon dioxide will be on the order of 1 percent of the excess carbon dioxide in the atmosphere. Within that 1 percent, the damages caused by a ton of emissions will be essentially constant.

Although a tax would be more efficient than a permit system for controlling greenhouse gas emissions (given flat marginal benefits, rising marginal costs, and high levels of uncertainty), a tax has a major political liability: it would induce large transfers of income from firms to the government. In fact, firms would end up paying far more in taxes than they spent on reducing emissions. For example, suppose that a particular firm was initially emitting Q tons of carbon dioxide and that its efficient abatement (where the marginal cost of abatement equaled the marginal benefit) was 20 percent. Under an efficient tax, T, the firm would eliminate $0.2Q$ tons of emissions at a cost no larger than $0.2QT$ (the firm would never pay more to abate its emissions than it would save in taxes, and it might pay much less if the marginal cost of abating the initial units of pollution was quite low). However, the firm would have to pay taxes on its remaining emissions, and its tax bill would be $0.8QT$, at least four times what it spent on abatement. This situation is illustrated in figure 5-4.

The political problem is not just that firms dislike paying taxes; it is that the transfers would be so much larger than the abatement costs that they would completely dominate the political debate. A firm that might be willing to pay $1 million to reduce its emissions by 20 percent would almost certainly be hostile to a policy that required it to pay $1 million plus an additional $4 million in taxes. The problem is not unique to climate change, and it is probably the primary reason that Pigouvian taxes have rarely been used to control environmental problems.

Incentive Compatibility

Economists are trained to worry about efficiency and to leave matters of equity and distribution to policymakers. With climate change, however, that dichotomy is untenable. There is no international agency to coerce countries to comply with an agreement that they find significantly inconsistent with their national interest, nor is there likely to be one in the foreseeable future. Greenhouse gas emissions originate throughout the world, and most countries eventually will need to participate in any reduction.

Figure 5-4. *Inframarginal Transfers under a Tax Policy*

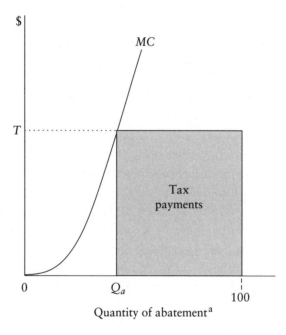

a. Percent of uncontrolled emissions.

However, a treaty that makes heavy demands on national sovereignty or that has significantly adverse effects on the distribution of income is unlikely to be ratified. In addition, countries that did ratify such an agreement would be likely to cease complying with it as soon as costs became substantial. A realistic climate policy therefore cannot rely on large international transfers to prevent individual countries from being made worse off. In the language of game theory, participation in a climate change agreement must be incentive-compatible for each country.

Unfortunately, much of the debate over the distributional aspects of climate change policy has focused on a different and far less practical question: which countries are ethically responsible for reducing climate change? Some observers argue that industrialized countries are obligated to do the most to avoid climate change because they have generated most of the greenhouse gases now in the atmosphere. Others argue that developing countries account for a large and growing share of emissions and that no climate policy will succeed without significant participation by the developing world. There is some truth in both of these positions, but neither is a realistic way to approach designing a policy that will have to be ratified by sovereign nations. Instead, the focus must be on developing a pragmatic policy that will allow all countries to make a firm commitment to cut emissions over time.

A Hybrid Approach

Although permits and taxes both have serious economic and political disadvantages when used alone, those problems can be overcome by a hybrid policy that combines the best elements of both.[3] For efficiency, the hybrid should act like an emissions tax at the margin: it should provide incentives for abatement of all emissions that can be cleaned up at low cost while allowing flexibility in total abatement if costs turn out to be high. For political viability, the hybrid should avoid unnecessarily large transfers and have the distributional flexibility of a permit system.

One policy that has these features is a modified permit system in which a fixed number of tradable, long-term emissions permits is supplemented by an elastic supply of short-term permits good for only one year (annual permits). Each country participating in the policy would be allowed to distribute a specified number of long-term emission permits, possibly in an amount equal to the country's 1990 emissions. The permits could be leased or traded without restriction, and each one would allow the holder to emit one ton of carbon per year. We refer to these as perpetual permits, although in principle they could have long but finite lives.[4] The permits could be given away, auctioned, or distributed in any other way the government of each country saw fit. Once distributed, the permits could be

---

3. A hybrid policy was first proposed by Roberts and Spence (1976). For more information about a hybrid approach to climate change policy, see McKibbin and Wilcoxen (1997a) and (1997b). This approach has also been endorsed by Kopp, Morgenstern, and Pizer (1997) and Victor (2001).

4. In earlier papers on this approach we have referred to these perpetual permits as endowments.

Figure 5-5. *Supply of Each Type of Permit for Use in a Given Year*

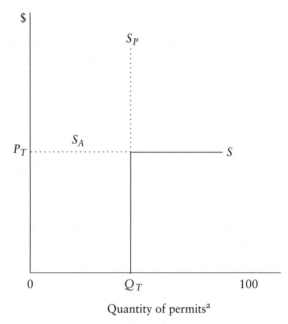

Quantity of permits[a]

a. Percent of uncontrolled emissions.

bought or sold among firms or even bought and retired by environmental groups. In addition, each government would be allowed to sell annual permits for a specified fee, say U.S. $10.[5] Firms within a country would be required to have a total number of emissions permits, in any combination of perpetual and annual permits, equal to the amount of emissions they produce in a year. In order to comply, firms without enough permits could buy or lease perpetual permits from other firms or buy annual permits from the government for the stated fee.

To see how the policy would work, consider the supply of permits available for use in any given year. On one hand, there will be an inelastic supply of Q_T perpetual permits, where Q_T is the number of such permits outstanding. This is shown in figure 5-5 by vertical line S_P. On the

5. Ten dollars per ton of carbon is equivalent to a tax of $6.50 per ton of coal or $1.40 per barrel of crude oil. See chapter 1.

Figure 5-6. *Abatement Costs and Demand for Permits*

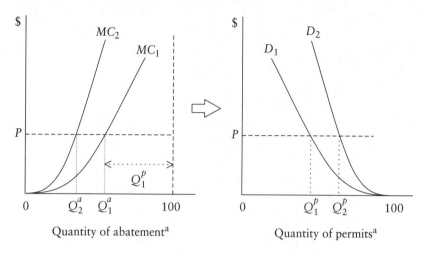

Quantity of abatement[a]   Quantity of permits[a]

a. Percent of uncontrolled emissions.

other hand, there also will be an elastic supply of annual permits available from the government at price P_T, which is specified in the policy. This is shown by horizontal line S_A in the figure. The total supply of permits is the horizontal sum of S_P and S_A, shown by the dark line in the figure.

The demand for permits will be determined by the marginal cost of abating emissions, as shown in figure 5-6. The left panel shows two hypothetical marginal cost curves for abatement: a low-cost curve, MC_1, and a higher-cost curve, MC_2. The right panel shows the corresponding demands for permits: D_1 and D_2. If costs are relatively low, as shown by MC_1, a given permit price P will induce Q_1^a units of abatement because those units can be eliminated for less than the price of a permit. As a result, firms will demand Q_1^p permits to cover their remaining emissions. If costs are relatively high, as shown by MC_2, firms will abate only Q_2^a units and they will demand a larger number of permits. Thus a flat marginal abatement cost curve implies a flat permit demand curve, and a steep abatement cost curve implies a steep permit demand curve that is relatively far from the origin.

Figure 5-7 shows two possible points of market equilibrium that could result from combining the supply and demand curves for permits. If

Figure 5-7. *Market Equilibrium in Low- and High-Cost Cases*

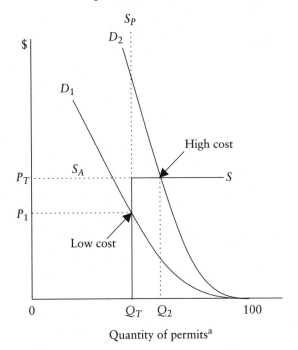

Quantity of permits[a]

a. Percent of uncontrolled emissions.

abatement costs turn out to be relatively low, permit demand will be low as well, as shown by curve D_1 in the diagram. In this case, the equilibrium permit price would be P_1, which is below the price of an annual permit, P_T. Only perpetual permits would be supplied, and emissions would be reduced to Q_T. If abatement costs turn out to be high, however, permit demand will be given by a curve like D_2. In that case, the price of a permit would be driven up to P_T and annual permits would begin to be sold. The final equilibrium price would be P_T, and the total number of permits demanded would be Q_2. Of the total, Q_T would be perpetual permits and Q_2–Q_T would be annual permits. The important thing to notice is that no matter how high the marginal costs turn out to be, the price of a permit would be capped at P_T.

Thus the hybrid has the key advantages of both tax and permit policies: it behaves like a tax policy at the margin, which is essential given the deep uncertainties inherent in climate change policy. Firms would have an incentive to reduce emissions whenever they could do so for less than P_T per ton. Because the total supply of permits would not be fixed, the policy would not guarantee precisely how much abatement would be done. Because it is a market-based instrument, however, it would ensure that any abatement would be done at minimum cost. Moreover, unless marginal costs of abatement are very low, the pattern of abatement will be efficient across countries as well as within each country. In particular, marginal abatement costs will be equalized in all countries where the price of a permit rises to P_T.

Like a permit policy, however, the hybrid avoids huge transfers of revenue to the government and creates a constituency (the owners of perpetual permits) that has a vested interest in maintaining the climate policy. The latter point is very important: the existence of a large group of permit owners would make it much harder for a government to renege on the policy. In addition, highly valuable perpetual permits provide a strong incentive for firms to investigate new methods or technologies to reduce emissions further.

6 *Implementing the Policy*

The mathematical analysis in chapter 5 might give the impression that the hybrid policy is a complicated and unfamiliar mechanism. In fact, nothing could be further from the truth. Stripped to its bare essentials, the main effect of the hybrid is to create a new asset—a perpetual permit—that would be traded and used very much like a conventional form of capital. When a firm needed to emit carbon dioxide, it would compare its prospective emissions against its stock of perpetual permits. If it had too few permits, it would have three choices: it could reduce its emissions until they matched its stock of permits; it could buy additional permits; or, if it expected the shortfall to be temporary, it could lease unused permits from other firms or buy annual permits from the government. Apart from the option to buy annual permits—which really just serves as a cap on leasing prices—those are exactly the choices that confront a firm when it has too little physical capital. In fact, even households face an analogous decision. When a household needs to transport passengers, it must compare the number of passengers to the capacity of its vehicles. If it cannot accommodate all of the passengers, the household can reduce the number of passengers; buy an additional vehicle; or, if it expects the problem to be temporary, rent a vehicle. The day-to-day operation of the hybrid policy, in other words, would be straightforward and familiar.

To implement the hybrid policy, a variety of institutional details would have to be addressed. Those details are the main subject of this chapter. None change the fundamental character of the policy, but they are an important part of our overall blueprint: they move a theoretical proposition one step closer toward practical reality.[1]

The Timing and Extent of International Participation

Climate change is a global problem with a very long time horizon. Eventually, all major countries will need to participate if a climate policy is to have a significant impact on the atmospheric concentration of greenhouse gases. At the same time, however, it is very unlikely that all countries will choose to participate simultaneously. Developing countries, for example, have repeatedly pointed out that industrialized countries are overwhelmingly responsible for greenhouse gas emissions to date and that those countries therefore should take the lead in reducing emissions. As a result, an international climate policy will need to cope with gradual accessions taking place over many years. Its design, in other words, must be suitable for use by a small group of initial participants, a large group of participants many years in the future, and all levels in between.

Because the hybrid policy is a harmonized system of domestic policies rather than a monolithic international policy, our approach has exactly the flexibility needed.[2] A country can participate simply by adopting the hybrid domestically, without any need for international negotiations. Moreover, accession by a new participant has no effect on the permit markets operating in other participating countries.

1. In addition, the details of implementation distinguish our policy from the Sky Trust hybrid proposed by Resources for the Future. The structure of Sky Trust is discussed by Kopp, Morgenstern, and Pizer (1997) and Kopp, Morgenstern, Pizer, and Toman (1999).

2. See chapter 7 for a more detailed discussion of the hybrid's robustness to accession by new participants.

The United States

The hybrid policy is decentralized enough to be launched by a single major country—the United States—acting unilaterally. Moreover, the U.S. government could take steps to establish domestic permit markets immediately, without waiting for a final international agreement to be negotiated; in fact, it would be in the nation's interest to do so, in order to allow the U.S. economy as much time as possible to adapt. A particularly attractive approach would be for the government to distribute the U.S. annex B allotment of perpetual permits immediately, even though a final treaty implementing the UNFCCC has yet to be negotiated. In doing so, however, the government would stipulate two conditions. First, firms would *not* be required to own permits for their emissions until an appropriate international climate agreement enters into force. Second, the permits would be guaranteed to be honored at their face value under any future climate agreement. The permits would be tradable without restriction.

On the surface, this suggestion might sound outrageous: we are proposing that the U.S. government hand out property rights for carbon emissions right now, long before it enacts any policy to restrict emissions and even before it is clear that there will be any restriction at all. The benefit, however, is very significant: distributing the permits now would allow financial markets to be far more effective at managing the risks of climate policy. Firms that are likely to be affected by climate regulations would suddenly have a new and powerful tool for managing their risk. For example, a firm that expected to have difficulty complying with a future constraint could reduce its risk by buying extra permits now as a hedge. If the policy is never enacted, the money paid for the permits will produce nothing, much like insurance premiums paid on policies when no claims are filed. On the other hand, if the policy *is* enacted, the firm will be partially protected against a potentially devastating event. At the other end of the spectrum, firms that expect to be able to comply easily with future restrictions will want to sell their excess permits to raise money now.

Pricing these permits would present a short-run challenge for financial markets: What is the value of an emissions permit when it is not certain when, if ever, carbon emissions will be regulated? However, that is precisely the type of problem that financial markets confront every day.

Within a very short time, an active market would develop with prices that reflect both the likelihood of a policy taking effect and its probable stringency.

Other Industrial Countries

If the United States took the lead in establishing the hybrid policy, other industrial countries would be presented with a choice. On one hand, they could ignore the U.S. initiative and continue attempting to implement the Kyoto Protocol. Doing so would not particularly harm the United States because the hybrid policy is decentralized, but it would do little to control climate change because the flaws in the Kyoto Protocol ultimately doom it to failure. On the other hand, some industrial countries might choose to follow the U.S. lead. The most likely candidates are precisely those whose participation is most important: countries with substantial emissions that would be reluctant to agree to the Kyoto Protocol's targets and timetables approach. As with the United States, it would be in the self-interest of those countries to begin building domestic permit markets.

Developing Countries

Any discussion of the role of developing countries in climate change policy should begin with the recognition of three facts: the participation of developing countries is crucial to the long-run success of any climate change policy; developing countries cannot be forced to participate in a climate change agreement that they find contrary to their own interests; and the transfers involved make it impossible for developed countries to pay developing countries to participate in anything but a minor way for a short period of time. Given these facts, the most a climate change policy can hope to do to facilitate participation of developing countries is to avoid being any more unattractive than necessary.

Among the alternatives, the hybrid policy has the best chance of being adopted by developing countries. Unlike the Kyoto Protocol, it requires little surrender of sovereignty. The perpetual permit markets in each country are independent of one another and are governed by each country's legal and institutional structures. In addition, accession by developing countries could be made smoother by assigning them initial allocations of perpetual permits that exceed their current emissions. Doing so would mean that their emissions could continue to grow for some time before the policy began having much effect. However, it would have an important political benefit: it would allow countries to commit themselves to full long-run participation in the agreement without imposing large burdens on their economies in the short run. It would allow firms plenty of time to adapt to the new regime while simultaneously sending them a clear signal, through the market for perpetual permits, that the cost of using fossil fuels would rise. In addition, it would allow time for the development of permit markets and other institutions before firms were actually required to comply.

It might be objected that reducing barriers is not enough, that a climate change policy must provide positive incentives for developing countries to participate or perhaps include penalties to punish them if they do not. That view, however, is both patronizing and naïve. Developing countries have far more at stake with respect to climate change than developed countries: their populations are larger, their economies are more vulnerable to the climate, and they are far less able to afford the measures that would be needed to mitigate climate change, such as sea walls, changes in agricultural practices, or even wider use of air conditioning. In fact, developing countries would be the principal beneficiaries of a climate change policy; to argue that they must be cajoled or coerced into acting in their own best interest is unwarranted condescension. The real issue is that developing countries face exactly the same trade-off as developed countries: on one hand, the risks of climate change are serious enough to justify modest steps to reduce the growth of emissions; on the other, too little is known to justify limiting emissions to a specific target regardless of cost. A policy such as ours, however, which fosters international cooperation in reducing emissions but does not impose rigid targets and timetables, will appeal to developed and developing countries alike. In addition, it does not rely on the politically unrealistic assumption that developed countries will be willing to devote substantial resources to bribing or coercing developing countries to participate.

Greenhouse Gases Other Than Carbon Dioxide

We focus on controlling carbon dioxide emissions because carbon dioxide is by far the most important greenhouse gas: it accounts for 60 percent of the increase in the atmosphere's heat-trapping capacity since the

Industrial Revolution. Moreover, carbon dioxide emissions are relatively easy to monitor because they result from the use of fossil fuels, and there tend to be few fuel producers in any given economy. However, the hybrid approach could be applied to other greenhouse gases as well. To control methane emissions, for example, countries would agree on an international allocation of perpetual methane permits and a trigger price at which governments could sell annual permits. The only real link needed between the policies for different gases would be to set the trigger prices in proportion to the global warming potential of each gas. For example, the trigger price for a metric ton of methane would be twenty-three times the trigger price of an equal amount of carbon dioxide.[3] Setting trigger prices in proportion to warming potentials would ensure that the overall reduction in radiative forcing achieved by each country was accomplished at the minimum possible cost.

Because the hybrid approach would control different gases with similar but separate policies, it allows controls on gases other than carbon dioxide to be added gradually over time. Since the only link between the policies would be the ratio of their trigger prices—which arises from physical properties of the gases rather than from negotiation—there would be no interaction between the policies and no need for all controls to be adopted simultaneously. Adding controls on methane, for example, would have little effect on the operation of a carbon dioxide policy already in place. Under the Kyoto Protocol, on the other hand, each country is given a single carbon-equivalent emissions allowance based on its 1990 emissions of specific gases. Adding a control on an additional gas would be difficult once the protocol had gone into effect: it would require adjustments to each country's emissions allowance and would have repercussions for permit prices around the world.

Issues Requiring International Negotiation

Although it is possible for countries to adopt different variations on the hybrid policy independently, over the long run it would reduce costs and

3. Table 2-2 lists the global warming potentials of various gases. The trigger price for methane would be twenty-three times that of carbon dioxide because one ton of methane traps twenty-three times as much heat as a ton of carbon dioxide.

improve the flow of information if the key features of the hybrid were harmonized across countries. The most important of these is the trigger price at which governments are allowed to sell annual permits; others include each country's allotment of perpetual permits; the treatment of sinks; and guidelines for monitoring. Each is discussed below.

The Trigger Price

The most critical feature of the hybrid policy that would benefit from harmonization is the price of annual emissions permits. If the trigger price differs across countries, emissions will be reduced but at an unnecessarily high cost. On the basis of our research with the G-Cubed multicountry model, we propose a price of US$10 (1990 prices) per ton of carbon as a starting point for international negotiations. A relatively low price is appropriate given current estimates of the benefits of controlling climate change and the wide range of uncertainties involved. An important advantage of the hybrid, however, is its flexibility: if new information becomes available, the trigger price can be renegotiated.

National Allocation of Perpetual Permits

The initial allocation of perpetual permits across countries is another aspect of the policy that would require international negotiation. However, the negotiations would be radically different from those over the Kyoto Protocol's annex B targets. Because the Kyoto Protocol requires countries to commit themselves to rigid emissions targets, negotiations inherently bring participants into conflict. Each country has an incentive to lobby for a loose target for itself, in order to keep its costs low and protect its economy, while simultaneously arguing for tight targets for other participants in order to reduce global emissions as much as possible. Each round of negotiations over targets for successive Kyoto commitment periods would be, in essence, a competition between national negotiating teams. Our approach, on the other hand, fosters cooperation rather than competition. One reason is that the policy requires countries to take mutually agreed actions—establishing a carbon dioxide permit market and selling annual permits, if needed, at the international trigger price—rather than requiring them to achieve specific individual emissions targets. The policy's defining characteristic, the trigger price, is uniform across countries, which eliminates the opportunity for beggar-thy-neighbor competition. Since the trigger price also provides an upper bound on compliance costs, it reduces the incentives for aggressive negotiation as well.

A second reason why our proposal would sharply reduce competition in negotiations is that it eliminates international conflicts over the initial allocation of perpetual permits. A country's allotment of perpetual permits no longer determines the amount of emissions it can generate. Rather, it determines the effect of the policy on the country's distribution of income: allowing a country to issue an additional perpetual permit will increase the wealth of permit owners and reduce government revenue from sales of annual permits. The difference between our proposal and the Kyoto Protocol, in other words, is that under our system changes in perpetual permit allotments result in transfers of income and wealth *within* countries while under the Kyoto Protocol they result in transfers *between* countries.

The Treatment of Sinks

An issue that has been important in the Kyoto negotiations is the role of greenhouse gas sinks: forests and other mechanisms that remove greenhouse gases from the atmosphere. Improvements in management often can increase a forest's net uptake of carbon dioxide. Countries with large forests have successfully argued that the treaty should increase their greenhouse gas emissions quotas by giving them credit for any enhancements they make in the amount of carbon dioxide absorbed by their sinks. Our proposal, however, would apply only to emissions and would not include a mechanism for incorporating sinks. In theory, there is a legitimate intellectual argument for counting sinks. They remove carbon from the atmosphere, and if they are excluded, low-cost sink activities that could offset high-cost carbon abatement activities might go unused. However, there are a host of technical difficulties in measuring sinks: quantifying the effect of new management practices on a forest's net carbon dioxide absorption is far more difficult than measuring the tons of coal removed from a mine. Moreover, negotiations to date have essentially reduced sinks to the role of a loophole that largely vitiates the protocol's constraint on emissions. Every sink credit that is allowed will be matched by an increase in the amount of greenhouse gases a country can emit. At the recent COP 6 meeting in Bonn, sink allowances were increased significantly to entice Japan and Canada to participate. The magnitude of sinks now allowed may even have effectively removed a cap on emissions in the first commitment period of Kyoto. Our policy excludes sinks, at least initially, to keep it as simple and transparent as possible.

Over time, it would be possible to extend the policy to include sinks. If measurement problems can be overcome, firms that enhance sinks could be given a corresponding number of annual emissions permits by their government. The firms could use the extra permits to cover part of their own emissions, or they could sell them on the open market. In effect, the addition of sinks would transfer revenue from the government to the private sector: the permits created by sink enhancement would reduce the number of annual permits sold by the government. Again there is a big difference between our approach and the Kyoto Protocol: under our approach the treatment of sinks is essentially a domestic issue, not a cause for international confrontation.

Monitoring and Compliance

One of the advantages of our proposal is that it shifts most monitoring and enforcement activities to the domestic level. Within each country, permit owners and the government will have strong incentives to monitor firms and bring violators into compliance: noncompliance hurts permit owners by reducing the value of their permits, and it deprives the government of revenue from selling annual permits. At the international level, however, monitoring and compliance would require only an annual statement from each country reporting its total emissions and its sales of annual permits. Annual permit sales would be itemized by buyer to increase the transparency and credibility of the report. A country would be out of compliance if it failed to file the report or had emissions exceeding the sum of its perpetual permit allocation and its annual permit sales. There would be no explicit penalty for noncompliance because no international institution exists (or is likely to be established in the foreseeable future) with the power to enforce one. However, the fact would become a matter of public record. Individual countries complying with the agreement would be free to take another country's noncompliance into account when negotiating bilateral or multilateral agreements on other matters.

Our proposal's relatively light requirement for international monitoring and enforcement has the additional benefit of making the agreement more robust than an international permit-trading system such as the Kyoto Protocol. Under our approach, noncompliance by one country affects its own citizens but has little effect on other countries. The noncompliant country might gain a small advantage in its terms of trade as a result of having lower energy prices, but that would be all. Under the Kyoto Protocol, in contrast, cheating by one country not only gives it a trade advantage, it also reduces worldwide demand for emissions permits, lowering the price of permits and with it the wealth of permit owners throughout the world. Without a strong international enforcement regime, an international permit trading system will be inherently unstable.

Domestic Implementation

Many details of the hybrid policy's implementation do not require international negotiation and would be left up to individual governments. These include the "point of control"—the specific activity for which permits would be required (generally either fuel production or use); the method used to distribute perpetual permits; the longevity of perpetual permits (which could be finite, if needed); mechanisms for monitoring firms and penalizing non-compliance; the structure and operation of permit markets; and integration of the hybrid with other climate policies. Each is discussed below.

Point of Control

Although the hybrid policy requires each ton of emissions to be matched by a corresponding permit, individual governments would be allowed to choose the specific stage in energy production at which to impose the requirement. The most obvious option would be the point of combustion: anyone burning fossil fuels would have to hold a corresponding number of permits. For some activities, such as electric power generation, that approach would work well. There are relatively few electric utilities, so monitoring costs would be low. Moreover, utilities would be able to take advantage of economies of scale in administering the new system.[4] For other activities, however, applying the policy at the point of consumption would be difficult or far more expensive. Gasoline consumption is a case in point: it would be impossible to monitor hundreds of millions of drivers to verify that each held an appropriate number of permits.

To keep administrative costs low, the best option for most governments is to apply the policy at the point where fossil fuels are extracted or imported, rather than at the point when they are consumed. Under that approach, compliance is monitored at the mouth of the mine, the wellhead, or the port of import, and firms must have a permit for every ton of carbon embodied in the fossil fuels they produce or import for domestic consumption. Fuels produced for export are exempt, although they do require a permit in the importing country, if that country is a party to the agreement. Because a relatively small number of large firms are involved in fossil fuel production, the policy's administrative and monitoring costs will be kept as low as possible.

Design and Operation of Permit Markets

One of the key advantages of our blueprint is that each country would manage its own domestic permit trading system using its own legal system and financial institutions. There is a strong case for cooperating across countries in terms of system design and especially in making expertise available for developing country participants. However, cooperation and information sharing would be the extent of the links across the markets. The markets themselves would need no other links because by international agreement the price of permits in the short term would be the same in each market (having risen to the trigger price); therefore, there would be no gains from trade.

Our proposal requires two new markets within each country. The first would be for ownership of perpetual permits. It would allow firms and individual investors to buy and sell permits, and it would operate much

---

4. In order to comply with the policy, a firm might need to hire additional managers or put new accounting controls in place. Once it had done so, however, its administrative costs would be largely independent of the amount of fuel it used.

like a market for very long term bonds. The price of a permit would be determined by market forces and would fluctuate freely in response to new information and changes in expectations about the future. The quantity of permits would be fixed by the country's internationally agreed allotment. Augmenting the ownership market would be a second market in which firms and individuals could buy and sell annual leases of perpetual permits. It would function as a spot market, allowing firms and permit owners additional flexibility in responding to short-term events. For example, a firm having unusually low emissions one year could lease its excess permits to a firm whose emissions that year were unusually high. The trigger price would function as a cap on the spot market price: a firm would never pay more to lease a perpetual permit for a year than it would have to pay to buy an annual permit from the government. The ownership and spot markets would be linked by arbitrage: traders would drive the price of owning a perpetual permit up or down until it equaled the expected present value of the stream of lease payments that could be earned on it.

Once the two fundamental permit markets were established, related markets would be likely to arise spontaneously. Financial institutions could use the permit markets to create derivative securities, such as call and put options, that would improve the efficiency of risk-sharing between firms and investors. For example, a risk-averse electric utility could hedge against future increases in the trigger price by buying call options. The options would entitle it to buy perpetual permits at a stipulated price during a specified period. If the trigger price were to rise sharply, the utility could exercise the options in order to increase its holdings of perpetual permits. On the other side of the transaction would be investors with a higher tolerance for risk (or with highly diversified portfolios) or firms expecting to reduce their future emissions at low cost. In effect, the transaction would allow the risk-averse utility to buy insurance against a sharp increase in the trigger price.

Implementing these markets would be straightforward, especially in developed countries. All of the new markets—those for derivatives as well as those for the underlying assets—are similar to markets that already exist. Moreover, the markets would produce considerable benefits: in addition to their role in providing a form of insurance, they would provide concrete incentives for households and firms to adjust their current behavior in anticipation of future climate policy. For example, if new scientific information became available suggesting that climate change would be more severe than expected, investors and firms would realize right away that the trigger price was likely to be raised—even though it might take governments months or years to agree on the actual change. The price of perpetual permits would be bid up immediately, stimulating energy conservation and emissions reductions.

Longevity of Permits

Another decision for individual governments is whether to issue true perpetual permits or to issue a rolling sequence of permits with long but finite lifetimes. Either would be compatible with our proposal as long as the total number of simultaneously long-lived permits did not exceed the country's internationally negotiated allocation. Permits with finite lives would entitle their owners to emit one ton of carbon each year until a specified date in the future; for example, a permit might be valid until 2050. As the expiration date of the permit approached, the government would issue a replacement permit valid during the subsequent period, say 2051 to 2100.

From a government's perspective, permits with expiration dates have two advantages. First, they make it substantially easier for the government to tighten the country's emissions constraint in the future. If the permits have infinite lives, the principal way to reduce emissions below the amount of the country's initial allotment of permits would be for the government to buy permits from the private sector and retire them from circulation. Such a policy could be very expensive. If permits had expiration dates, however, the government could tighten the constraint simply by issuing fewer permits in subsequent periods. The second advantage of expiration dates is that the government would be able to change the distributional impact of the policy by adjusting the allocation of permits at each renewal date. For instance, the first round of permits might be given to energy producers as a form of grandfathering while subsequent rounds would given to households, which would sell them to firms, keeping the revenue as partial compensation for higher energy prices.

Carrying the notion of expiration dates one step further, a government could even issue a range of long-term permits with a variety of expiration dates, much the way governments now issue bonds. For example, a country with an allowance of 100 long-term permits might chose to issue twenty of them as perpetual permits, forty as permits expiring in fifty years, and the remaining forty as permits expiring in twenty years. In essence, this approach would create a family of assets with a term structure of expiration dates.[5]

Perpetual permits, however, have advantages of their own. First, they establish clear, unambiguous property rights, reducing the uncertainty faced by firms and facilitating long-term planning and investment. Second, they establish strong constituencies with a vested interest in maintaining the policy. Third, they create a single large permit market, improving the efficiency and transparency of trading; permits with varying expiration dates would be separate goods and would trade in smaller markets.[6] Fourth, a one-time distribution of permits would reduce the overall cost of the policy by eliminating the incentive for firms to pour money into lobbying for higher future allocations of permits. In our view, the advantages of perpetual permits far outweigh those of permits with expiration dates. Our policy, however, would allow individual governments to come to their own conclusions.

Mechanisms for Allocating Permits

A successful climate policy must allow each participating country wide latitude in how it addresses distributional issues within its borders. For example, domestic political considerations might make it essential for a government to compensate parties hurt by the policy or at least to provide some sort of transitional relief, such as grandfathering a portion of each

5. Nicholas Gruen and Geoff Francis have made similar suggestions to us along these lines.

6. When applied to markets, the term "transparent" means that all parties to a transaction can easily understand exactly what is being exchanged. Transparency is desirable because it improves the efficiency of markets by reducing the cost of transactions. Many energy markets lack transparency because transactions are governed by complex contracts that make prices and payments contingent on the outcomes of an array of uncertain future events.

existing firm's emissions. An ideal agreement would be designed to allow countries this kind of distributional flexibility without undermining the international transparency of the policy. The following sections discuss the advantages and disadvantages of several mechanisms that might be used to allocate the permits.

GRANDFATHERING. There are a number of reasons why a government might want to grandfather some of its emissions-intensive industries by giving them a portion of the country's perpetual permits. One reason is equity: the shareholders of those firms will suffer a capital loss from raising the price of carbon because it is difficult and expensive to reallocate capital from one industry to another and it is costly to retrain workers. It is inevitable that effective greenhouse policies will cause a reduction in the use of carbon. Thus the allocation of perpetual permits could be used to compensate owners of the industries most affected. A second reason is more pragmatic: distributing permits for free builds a constituency with a vested interest in having the policy succeed.

In addition, equity concerns might prompt a government to distribute a portion of the permits to workers in energy-intensive industries or to households that purchase energy or energy-intensive goods and that would face higher prices as a result of the policy. The recipients would not use the permits themselves but would be able to sell or lease them to firms. The revenue they would receive would compensate them, at least in part, for the changes in the energy sector.

AUCTIONING. Another alternative would be to distribute the permits by auction.[7] The advantage of an auction would be that it would generate government revenue that could be used to reduce tax distortions elsewhere in the economy. The disadvantage would be that it would essentially eliminate the difference between the hybrid policy and a plain carbon tax. Moreover, it would do so in a manner that firms would find particularly unpleasant.

Unless the trigger price was set so high that no annual permits were likely to be sold (in which case a plain permit policy would have been sufficient), the price of a perpetual permit sold at auction would rise to the

---

7. The hybrid policy advocated by Kopp, Morgenstern, Pizer, and Toman (1999) uses auctions to allocate emissions permits.

present value of buying an equivalent series of annual permits. From the point of view of a firm bidding on a perpetual permit, therefore, the policy would look exactly like a carbon tax set at the trigger price, with one important difference: the present value of all future taxes would have to be paid up front, at the time the permit was purchased.

To make this point concrete, consider the following example. Suppose that the trigger price is set to $10 per ton of carbon and that most firms believe that real permit prices will rise to that level every year (that is, they believe that at least a few annual permits will be sold every year). A perpetual permit will allow its owner to avoid paying $10 every year. At a real interest rate of 5 percent, the present value of a stream of annual $10 payments is $200. As a result, the auction price of a perpetual permit would rise to $200—exactly the present value of paying a $10 per ton carbon tax every year forever. Capitalization of future trigger prices would be less severe for permits with shorter life spans, but the equivalence between auctioned permits and emissions taxes would still hold.

Monitoring and Compliance

An important strength of our proposal is that it creates internal incentives in each country for monitoring and enforcing the policy. On one hand, the government would have an incentive to monitor firms and enforce compliance because doing so would increase the number of annual permits it would sell, allowing it to raise revenue that could be used to reduce other taxes or provide government services. On the other hand, firms and households that own perpetual permits would also have a strong financial incentive to make sure that firms comply with the policy: compliance preserves the asset value of perpetual permits. With poor monitoring and enforcement, firms will buy fewer permits than they actually need, causing the market price of a permit to be low and reducing the wealth of permit owners. In the worst case, when no firms comply, the price of a perpetual permit would fall to nearly zero. Permit owners, as a result, can be expected to lobby their own governments vigorously in support of enforcing the policy.

Combining the Hybrid with Other Policies

In the long run, it is very likely that a portfolio of policies will be used to address climate change. In addition to a broad market-based instrument such as the hybrid policy, other measures that might be adopted include reductions in existing fuel subsidies, subsidies for research and development (particularly for energy technology), voluntary emissions reduction agreements between firms and the government, information campaigns, demand-side management, and technology and performance standards. Each of these could be combined with the hybrid, but none of them could replace it. In fact, without a price-based instrument like the hybrid, many of these policies would be counterproductive. Subsidized research and development, in particular, would have the effect of *reducing* energy prices, thus tending to increase energy consumption and greenhouse gas emissions. Using the hybrid policy in combination with a research subsidy would offset this effect.

Evolution of the Policy over Time

The enormous uncertainties and long time frames involved in climate change make it inevitable that any climate policy will need to be revised. A final issue in the practical implementation of our blueprint is to anticipate how it would evolve over time as more information becomes available.

The key attribute of our policy requiring revision is the trigger price – the price at which governments are allowed to sell annual permits. If scientific evidence accumulates showing that severe effects of climate change are more likely than currently thought, it would be necessary to reduce emissions by raising the trigger price. On the other hand, if current uncertainties remain unresolved, or if firms discover new technologies or other unexpected methods of reducing emissions at low cost, the trigger price might be left unchanged.[8]

In order to ensure that the trigger price remains appropriate over time, it would be necessary to continue periodic COP meetings. However, the frequency of meetings could drop substantially in light of the gradual accumulation of scientific knowledge. Rather than meeting every year or two, as is now the case, COP meetings could be held once a decade. At

---

8. Improvements in technology would not require reductions in the trigger price: it puts an upper bound on compliance costs, but it would not prevent the lease value of a perpetual permit (and with it, the purchase price of a permit) from falling if compliance became inexpensive.

the same time, the structure of our policy would make COP meetings more productive and less acrimonious. The only issue for negotiation would be a single number: the trigger price. Negotiations would be transparent and facilitated by the fact that the trigger price allows each country to predict its worst-case compliance cost easily. Under the Kyoto Protocol, in contrast, every country's emissions allocation would have to be renegotiated and each participant would once again be faced with having to agree to achieve its commitment regardless of cost.

The only situation that would require a revision to our policy beyond a change in the trigger price would be an acute increase in the expected damages from climate change. In that case it could become necessary to reduce each country's emissions to a level below its agreed allotment of perpetual permits. Even here, however, international negotiations would be less confrontational than they would be under the Kyoto Protocol or another targets and timetables approach. As noted above, a reduction in a country's perpetual permit allotment shifts revenue within the country—between the government and the private sector—but generally not between countries.

7 Comparing the Hybrid Policy with Alternatives

The fundamental strength of the policy that we propose in this book is that it is a hybrid of the two best-known market-based instruments for controlling pollution—an emissions tax and a tradable permit system. It avoids the pitfalls of each but retains the main benefits of both. This can be seen in table 7-1, which compares the hybrid with the most prominent examples of tax and permit policies: a harmonized carbon tax[1] and the Kyoto Protocol.[2]

All three policies have some features in common. Because they are market based, each policy minimizes the

---

1. A harmonized carbon tax has been proposed by Nordhaus. Under such a policy, each participating country would adopt a domestic carbon tax equal in value to a stated amount in a single currency (for example, US$10).

2. In this chapter we compare the hybrid with two broad classes of market-based policies: taxes and tradable permits. The relationship between the hybrid and a range of non-market-based policies is discussed in chapter 6.

Table 7-1. *Comparison of Key Attributes of Market-Based Climate Policies*

Attribute	Kyoto Protocol	Carbon tax	Hybrid policy
Common attributes			
Minimizes abatement costs within each country	yes	yes	yes
Encourages energy conservation and innovation	yes	yes	yes
Guarantees that benefits are greater than costs	no	no	no
Attributes of a tax-based approach			
Economically efficient response to uncertainty	no	yes	yes
Explicit upper bound on compliance costs	no	yes	yes
Avoids large international transfers of wealth	no	yes	yes
Provides incentives for domestic enforcement	no	yes	yes
Does not need strong international enforcement	no	yes	yes
Robust to accession or withdrawal of participants	no	yes	yes
Low disincentives for developing countries	no	yes	yes
Integrates easily into electric utility price regulation	no	yes	no
Attributes of a permit-based approach			
Flexibility in domestic distributional effects	yes	no	yes
Does not requires large transfers to the government	yes	no	yes
Easy to implement transition relief	yes	no	yes
Creates constituencies for enforcement	yes	no	yes
Guarantees a given reduction in emissions	yes	no	no
Other attributes			
Minimizes abatement costs across countries	depends	yes	depends
Short-run incentives for developing countries	CDM[a]	no	no

a. Clean development mechanism.

domestic cost of abatement done within each country.[3] They also raise the price of carbon emissions, encouraging conservation and stimulating innovation in energy-saving technology. In addition, the enormous uncertainties in climate change mean that none of the policies is guaranteed to have benefits larger than its costs.

3. For the purposes of this comparison we are assuming that each of the policies would be implemented efficiently. In practice, administrative costs, restrictions on permit trading, transactions costs, or other imperfections could keep costs from being driven to their absolute minimum.

Strengths Shared with a Harmonized Carbon Tax

The hybrid policy and a carbon tax also share many advantages not possessed by the Kyoto Protocol. By far the most important is that the policies are economically appropriate given the uncertainties discussed in chapter 2. Both ensure that action would be taken to reduce greenhouse gas emissions, but unlike the Kyoto Protocol, they limit the marginal cost of abatement and do not obligate participating countries to hit a particular emissions target regardless of the cost. Not only is that important for economic efficiency, it also is an important political benefit. When a climate treaty eventually comes before the U.S. Senate for ratification, a key question that will arise in the debate over ratification is whether the costs are guaranteed not to be excessive. Under both a carbon tax and the hybrid policy, the answer is "Yes." As a result, governments of the countries that matter most for climate change—those with substantial greenhouse gas emissions—could ratify a treaty based on one of these policies without an open-ended surrender of sovereignty.

Both an emissions tax and the hybrid policy also have several advantages arising from the fact that they do not involve international permit trading. Most important, they do not involve large international transfers of wealth, which eliminates a major political obstacle that the Kyoto Protocol would face during the ratification debate.[4] In addition, they avoid the potentially serious effects that international permit trading would have on exchange rates and trade balances.

Both policies also would generate government revenue that would be available for a variety of purposes: to reduce budget deficits, lower personal income taxes, or shore up social insurance programs. This would give domestic governments an incentive to monitor and enforce the agreement within their borders. As a result, the policy could be implemented with a much weaker international enforcement mechanism than the

---

4. This advantage is not shared by an international emissions tax, a policy that differs more sharply from a harmonized emissions tax than the similarity of the names would suggest. A harmonized carbon tax would be levied by each participating government on its own citizens, and the governments would be free to use the revenue as they saw fit. An international emissions tax would be levied by an international agency and the proceeds would not necessarily be returned to national governments.

Kyoto Protocol would require. The main role for an international agency would be monitoring and publicizing national activities. A related advantage is that there would be no need for an international liability mechanism to determine what to do when an entity sells permits without making corresponding reductions in its emissions; such problems would be handled by the existing legal system in each country.

Another strength shared by the hybrid and a carbon tax is that it would be easy to add countries to the system over time: those interested in joining would only have to adopt the policy domestically; no international negotiations would be required. That flexibility is crucial, because it is clear from the history of climate negotiations that only a few countries would now be willing to implement a significant global warming treaty in the near future. Furthermore, countries could withdraw from the system without debasing the value of the permits in countries that continued to participate. Under a pure system of internationally tradable permits, the addition or withdrawal of any country could cause large swings in the price of permits.

Finally, a carbon tax or the hybrid policy would also be more likely than the Kyoto Protocol to elicit significant long-term participation from developing countries. The main reason is identical to that for developed nations: a tax or the hybrid do not require participating countries to agree to rigid limits on their greenhouse gas emissions. The targets and timetables approach of the protocol is especially unattractive to the governments of developing countries, which are wary of choking off future economic growth. As a group, developing countries have repeatedly emphasized in UNFCCC and COP negotiations that they are unwilling to make emissions commitments like those in the protocol's annex B. Developing countries may participate in the protocol in a limited way through the clean development mechanism, but there is no long-term incentive for them to accede to the full agreement. The main advantage of a tax or the hybrid policy is not so much that either one offers developing countries a strong incentive to participate; rather, they do much less to discourage participation.

Strengths Shared with an International Permit System

The hybrid policy also shares several advantages of a tradable permit system. Most important, both policies allow domestic governments great flexibility in managing distributional effects. One possibility would be to auction the permits, but while doing so would raise considerable revenue for the government, it would eliminate the distributional and political advantages of a permit system over an equivalent tax. Another alternative would be to distribute the permits for free to firms in proportion to their historical greenhouse gas emissions.[5] In this case, the permits would amount essentially to a form of grandfathering, which could be a very important political benefit. Firms would be much less likely to oppose the policy if it exempts a portion of their emissions. In addition, the permits would provide a form of transition relief to firms that had to make major changes in their activities in the wake of the policy. For example, if higher energy prices cause a firm to stop making a particular product, the firm will be able to cover some if its losses by selling the emissions permits it no longer needs. Other policies are possible as well, such as giving some or all of the permits to households, which would be able to sell them to emissions-producing firms and thereby receive partial compensation for the higher energy prices they would face. Another option would be to sell the permits at a price below their market value, although that would require a rationing mechanism because the demand for permits would exceed the supply.

As long as the permits are not auctioned, either the Kyoto Protocol or the hybrid policy would have a substantial political advantage over a carbon tax: both policies avoid the large transfers of income to the government that would occur under a tax. Moreover, any transfers that did occur would be between firms or between firms and households rather than between the private sector and the government.

Another advantage shared by the hybrid and the protocol is that the owners of long-term emissions permits will have a vested interest in the success of the policy. Permit owners will have strong economic incentives to oppose any weakening of the protocol because that would reduce the market value of their permits. This is a particularly important feature because a climate change policy will have to remain in place indefinitely in order to keep atmospheric concentrations of greenhouse gases from rising. Without strong, influential constituencies it is likely that a climate

---

5. For example, each firm could be given a number of permits equal to 90 percent of its 1990 emissions.

policy would be abrogated during a future economic downturn or when compliance costs became burdensome.

A final strength of both approaches is that they could be used to create incentives for taking early action to reduce greenhouse gas emissions. All that would be needed would be to distribute the policy's permits immediately even though emissions limits would not be binding for several years. This would allow firms to begin trading now in anticipation of future needs. A firm that could reduce its emissions easily, for example, could do so in order to sell its permits. Moreover, distributing the permits immediately would allow the use of sophisticated financial instruments to help firms cope with uncertainty. For example, a firm expecting to have difficulty reducing its emissions could hedge its position by buying call options on additional permits. By doing so, it would be able to reduce its risk.

The potential gains from harnessing financial markets to manage risk are so large, in fact, that governments should unilaterally begin to issue permits immediately even though no binding climate change agreement has been ratified. The effect of the permits would depend explicitly on the adoption of an international climate change agreement: if an agreement is adopted, each permit will entitle the owner to emit an amount of greenhouse gases equivalent to one ton of carbon; if no agreement is adopted, greenhouse gas emissions would be unconstrained and the permits would do nothing. The market value of such a permit would rise and fall with changes in the expected values of an array of important unknowns: the likelihood of an agreement; the date when the agreement would come into force; the rate of growth of energy consumption; technological innovation in alternative energy technologies and noncarbon fuel; new discoveries of fossil fuels; and many others. Many of these are highly uncertain, but they are exactly the sort of uncertainties that financial markets confront every day. Issuing such permits would help firms and hence the economy manage the risks posed by potential future climate agreements.

What the Hybrid Policy Does Not Do

The hybrid policy combines most of the advantages of an emissions tax and a tradable permit system. There are, however, a few features of one or the other that it does not have. Unlike a harmonized carbon tax, the hybrid does not guarantee that marginal abatement costs will be equal in every country and hence does not guarantee that abatement costs will be minimized internationally. Nevertheless, cost minimization is very likely. Once the policy is in place, permit prices will be equalized at the international trigger price in all countries where emissions exceed the country's allotment of long-term permits. The only circumstance in which prices would not be equalized is that in which the emissions of one or more countries are at or below the country's long-term permit allotment. In that case, permit prices would be lower in that country than elsewhere.

The only major country in which this scenario is likely is Russia, whose emissions currently are far below the country's annex B allotment. Because Russia has surplus allowances, however, the difference in permit prices between it and other annex B countries would not reflect differences in the marginal cost of abatement as much as it would a loosening of the environmental constraint. Permit trading with Russia would lower total abatement costs because total annex B emissions could rise by several hundred million metric tons of carbon, not because of an improvement in the allocation of a fixed amount of abatement.

Moreover, it is important to note that the Kyoto Protocol does not guarantee that permit prices will be equalized across countries either. Permits originating in different countries will trade at different prices because of variations in the risk of noncompliance. As a practical matter, any international inefficiencies arising under either the hybrid policy or the Kyoto Protocol are likely to be small, and the political costs of eliminating them through trading may be high. Appendix A examines this trade-off in more detail.

One potential advantage possessed by the Kyoto Protocol but not by the hybrid policy or an emissions tax is a short-run incentive for the participation of developing countries. Neither of the latter policies has a feature equivalent to the Kyoto Protocol's clean development mechanism (CDM). For a variety of reasons, it is not clear how much abatement this mechanism would actually achieve. Many institutional details remain to be resolved, and CDM projects would be attractive only if permit prices were relatively high in the annex B trading regime. If the annex B trading system came into force today, for example, permit prices would be very low because emissions from annex B countries are substantially below annex B commitments.[6] That will change as annex B economies grow over time, but CDM projects are unlikely to be very attractive for quite a while. It would be overstating the case for the hybrid policy to assume that use of the mechanism will be zero, but it is not likely to be very large. More important in the long run will be encouraging developing countries to participate fully in an international climate agreement, and the targets and timetables design of the Kyoto Protocol is a very substantial disincentive.

Both the hybrid policy and the Kyoto Protocol lack a potentially important characteristic possessed by a carbon tax: easy integration into the existing system of electric utility regulation. Public utility commissions would almost certainly treat a carbon tax as analogous to a fuel price increase and allow electric utilities to pass it on in the form of higher electricity prices. As a result, customers would reduce their electricity consumption when they could do so at low cost; fuel use and greenhouse gas emissions then would fall. It is less clear how regulators would treat permits. Ideally, electricity prices would increase in proportion to the market price of an annual emissions permit. If utilities had to buy annual permits to cover all of their emissions, there would be no problem: regulators would almost certainly allow the cost to be reflected in electricity prices. Most utilities, however, would own many perpetual permits. For these to have the appropriate effect on electricity prices, public utility commissions would need to allow utilities to count as part of their costs the amount they forgo by not leasing their permits to other firms. This approach is reasonably straightforward to implement, but it has not received much attention in the literature.

Finally, neither the hybrid policy nor a carbon tax guarantees that greenhouse gas emissions will be limited to a particular value. The Kyoto Protocol, in contrast, restricts annex B emissions to about 95 percent of their 1990 value. Under the hybrid, emissions could fall to that level if abatement is relatively cheap; on the other hand, if abatement is expensive, emissions will be higher than that. However, the lack of a rigid target is not an accident or a defect in the policy: rather, it is the policy's sin-

---

6. This is true because of the contraction since 1990 in the economies of Russia and other parts of the former Soviet Union.

gle greatest strength. As discussed throughout this book, the rigid targets and timetables approach is economically inappropriate and politically unrealistic given the uncertainties involved in climate change. The hybrid policy, on the other hand, is more efficient and more realistic because it ensures that emissions will be reduced but does not require countries to agree to achieve a given emissions target regardless of cost.

Summary

In summary, the hybrid policy combines all of the important strengths of both an emissions tax and a tradable permit policy. Like a tax, it is economically efficient given the uncertainties of climate change. Like a permit policy, it can easily be adjusted to achieve a variety of distributional effects. As a result, it is far more realistic than either of the alternatives. Because it does not require signatories to commit to achieving a specific emissions target regardless of cost, it is more likely to be ratified than the Kyoto Protocol. Because its distributional effects would be much more acceptable, its political prospects are much better than those of a carbon tax. Overall, a hybrid policy is an efficient and practical approach to climate change.

8 | *How to Proceed from Here*

C limate change poses two challenges for policymakers: it is fraught with enormous uncertainties that
are unlikely to be resolved for decades; and the distributional effects of any policy will be critical to its prospects
for ratification. Both need to be taken into account during
the policy's design. In this book we show that it is possible
to design a system that can deal directly with these issues.
The appropriate policy is a hybrid combining the best features of an emissions tax with the distributional advantages
of a permit system. It would be efficient, practical, and
politically realistic.

Unfortunately, international negotiations to date have
produced a very different, deeply flawed policy, the Kyoto
Protocol. The protocol fails to address the uncertainties
inherent in climate change; as a result, it is inefficient and
politically unrealistic. It never had any chance of ratification by the U.S. Senate, and it was ultimately rejected by
the Bush administration. Other countries with large greenhouse gas emissions, including Russia, Japan, and Canada,
continued to participate in negotiations only because the

protocol's emissions targets were continually relaxed through expansions in the allowances for sinks. Consequently the protocol's emissions targets are now so loose that the treaty will do little or nothing to restrain emissions during the first commitment period, 2008–12. Emissions commitments for subsequent periods have yet to be negotiated, but there is nothing in the history of climate negotiations to suggest that they will be tightened significantly. Moreover, if future targets do become tight, the protocol is likely to collapse because it lacks an effective compliance mechanism. Countries that found it difficult to comply with the protocol would simply withdraw from it.

International climate negotiations have now reached a critical juncture. The path of least resistance is to continue negotiations over implementation of the Kyoto Protocol. Many governments will find that path politically attractive because the protocol's rigid targets create the illusion that greenhouse gas emissions will be substantially reduced. However, the protocol's generous sink allowances and vast supply of unused permits from Russia mean that emissions could continue to rise for many years before the constraint became binding. Continuing down this path, in other words, allows current policymakers to postpone real action for years, allowing emissions to grow unchecked and passing the climate problem on to their successors.

Toward a More Realistic Approach to Climate Policy

The alternative approach, which we advocate in this book, is to acknowledge that the targets and timetables approach embodied by the Kyoto Protocol is fatally flawed and will never achieve the goals of the UNFCCC. Doing so would be a significant step forward because it would allow policymakers to address climate change in a serious but far more practical manner. Adopting an approach based on the principles we have outlined here would produce an international policy regime with many advantages over the Kyoto Protocol: it would have a realistic chance of ratification in countries with large carbon emissions; it would be much less unattractive to developing countries; it would have the flexibility to include all countries over time; and it would be politically sustainable, with the institutions and mechanisms needed to support a serious climate change policy for decades into the future.

Although the targets and timetables approach must be discarded, it would not be necessary to begin negotiations from scratch. Much of the hard work that went into negotiating the Kyoto Protocol produced results that could be incorporated into an alternative agreement. For example, the Kyoto annex B targets would be a natural basis for the allotment of perpetual permits to annex B countries. Allotments of perpetual permits to developing countries would still need to be negotiated, but that could be left for the future because the policy regime would not be disrupted by future accessions. In addition, many institutions already established would continue to have valuable roles to perform, such as developing international standards and techniques for monitoring and measuring emissions and sinks or helping individual countries implement the results of research done elsewhere. These institutions would foster international cooperation, which in the long run would be essential to operating an effective climate change system. Our approach encourages cooperation far more than the Kyoto Protocol does because it does not take the form of a zero-sum game in which countries must compete for a rigidly limited supply of emissions rights.

Early Action and Financial Markets

The decentralized nature of our approach means that individual countries could begin creating domestic permit markets immediately, without waiting for a final international agreement to be negotiated. In fact, it would be in each country's interest to do so, in order to allow its economy as much time as possible to adapt and its financial markets to help manage the risks of climate policy. One approach would be for a country to distribute its annex B allotment of perpetual permits immediately, even though a final treaty implementing the UNFCCC has yet to be negotiated. The government would stipulate that emissions permits would *not* be required unless an international or domestic decision is made to reduce carbon emissions and that the permits would be honored at their face value if a carbon constraint was eventually imposed. The permits would be tradable without restriction.

This approach does not reduce any of the deep uncertainties about climate change: it still will take decades to obtain more complete information about the risks and costs of greenhouse gas emissions. It also does little to reduce uncertainty about whether emissions eventually will be regulated. However, it would be nearly costless for the government to carry out, and it would confer a single, but critical, benefit on the economy: firms that are likely to be affected by climate regulations suddenly will have a new and powerful tool for managing their risk. For example, a firm that expected to have difficulty complying with a future constraint could reduce its risk by buying extra permits now as a hedge. If the policy is never enacted, the money paid for the permits will produce nothing, much like insurance premiums paid on policies when no claims are filed. On the other hand, if the policy *is* enacted, the firm will be partially protected against a potentially devastating event. At the other end of the spectrum, firms that expected to be able to comply easily with future restrictions could sell their excess permits to raise money now.

Of course, pricing these permits would present a short-run challenge for financial markets: what is the value of an emissions permit when it is not certain when, if ever, carbon emissions will be regulated? However, that is precisely the type of problem that financial markets confront every day—it arises, for example, when investors decide how respond to a firm's news about a new patent it has received on a product that has never been produced. Within a very short time, an active market would develop with prices that reflected both the likelihood of a policy taking effect and its probable stringency.

Time for a Change

Now is clearly the time for changing course on international climate change policy. The fundamental flaws in the Kyoto Protocol were apparent even before the agreement was signed; they now are acknowledged by all but the protocol's most devoted partisans.[1] There is little point in continuing to negotiate a treaty that does not include the United States and developing countries. Changing course will not be easy. The 2001 Marrakesh COP meeting showed that many participants believe that the protocol's lack of participation by major emitters can somehow be overcome

---

1. See McKibbin and Wilcoxen (1997a and b) for a critique of the draft proposal that formed the basis for the Kyoto Protocol.

with enough negotiating momentum. A far more productive approach, however, would be to begin addressing the protocol's flaws as soon as possible.

In this book we offer a way forward that acknowledges the deep uncertainties inherent in climate change policy and approaches them in a realistic and practical way. The advantages of the hybrid approach are so compelling that it is safe to say that sooner or later, it will be adopted in some form. Whether it is adopted now or only after more time and resources are poured into the doomed Kyoto Protocol is up to policy-makers. However, the longer the world waits for an effective and flexible climate change policy, the worse the climate problem becomes and the more expensive it will be to address it.

International Permit Trading: Efficiency and Transfers

Allowing emissions permits to be traded internationally can improve economic efficiency, but it can also lead to large international transfers of wealth. Unfortunately, these features go hand in hand: the conditions under which international trading is most important for efficiency are also the ones that lead to the largest transfers. Since large international transfers will reduce the political viability of a proposed policy—and even the *potential* for large transfers is likely to have a chilling effect—it is important to investigate whether the efficiency gains from trading are large enough to be worth the political and administrative problems trading would create. One way to approach this is to calculate how large the gains from trade are relative to the transfers involved. In this appendix we present a simple economic model that can be used to explore that question.

Suppose that a group of countries would like to lower its greenhouse gas emissions by Q^* tons and that the marginal cost of abatement for country i depends on the amount of abatement, q_i, that it does: $mc_i(q_i)$. Pareto efficiency requires that the allocation of abatement across countries be such that marginal costs are equal between all pairs of countries, i and j:

(1) $$mc_i(q_i^*) = mc_j(q_j^*),$$

where q_i^* and q_j^* are the efficient quantities of abatement in the two countries. In addition, total abatement must add up to Q^*:

(2) $$\sum_i q_i^* = Q^*.$$

For convenience, let the marginal cost under the efficient allocation of abatement be denoted by A:

(3) $$A = mc_i(q_i^*) = mc_j(q_j^*).$$

Now suppose that the marginal cost of eliminating a ton of emissions for country i when abatement is q_i can be written as follows:

(4) $$mc_i = A \left(\frac{q_i}{q_i^*} \right)^{\eta_i},$$

where η_i is an elasticity reflecting the rate at which marginal costs increase as abatement departs from its efficient level. When abatement is allocated efficiently, q_i is equal to q_i^* and mc_i will equal A. A 1 percent increase in q_i above q_i^* will raise marginal costs by approximately η_i percent.

Integrating (4) gives the total cost to country i of abating q_i units:

(5) $$tc_i = \frac{A}{(1 + \eta_i)} \left(\frac{q_i}{q_i^*} \right)^{\eta_i} q_i.$$

At the efficient allocation of abatement across countries, $q_i = q_i^*$ and (5) can be rewritten:

(6) $$tc_i^* = \frac{A q_i^*}{1 + \eta_i}.$$

When the cost elasticity η_i is very small, marginal abatement costs are nearly constant at A and tc_i is close to $A q_i^*$. When η_i is larger, the total cost of q_i^* declines (holding the marginal cost of q_i^* constant at A) because inframarginal units are cheaper to abate. Put another way, a larger η_i means a steeper marginal cost curve and hence a sharper reduction in marginal costs when q_i is below q_i^*. When η_i is 1.0, for example, tc_i drops to half of Aq_i^*.

Now suppose that nontradable emissions permits are handed out and that country i's allotment requires it to do an amount of abatement that differs from the efficient pattern by a percentage ε_i:

$$(7) \qquad q_i = q_i^*(1 + \varepsilon_i).$$

The sum of the q_i's remains at the efficient value Q^*. ε_i will be positive when country i has been given relatively few permits and therefore must abate more than its efficient amount; it will be negative in the opposite case. Because q_i differs from q_i^*, marginal costs will no longer be equal across countries. Inserting (7) into (4):

$$(8) \qquad mc_i = A(1 + \varepsilon_i)^{\eta_i}.$$

Total abatement costs in each country will differ from their efficient values as well, as can be seen by inserting (7) into (5), collecting terms, and then using (6) to write the actual costs in terms of the efficient value:

$$(9) \qquad tc_i^N = tc_i^*(1 + \varepsilon_i)^{1 + \eta_i}.$$

In contrast, if the permits were internationally tradable in a competitive market, the equilibrium price of a one-ton permit would be A and country i would buy $\varepsilon_i q_i^*$ permits from abroad (or sell permits, when ε_i is negative) at a total expenditure given by:

$$(10) \qquad tc_i^P = \varepsilon_i q_i^* A.$$

Using (6) allows this to be rewritten in terms of efficient total costs:

$$(11) \qquad tc_i^P = \varepsilon_i tc_i^*(1 + \eta_i).$$

This would allow it to move its level of abatement to q_i^*, and its total cost—including both abatement and permit purchases or sales—would become:

$$(12) \qquad tc_i^T = tc_i^* + tc_i^P.$$

Inserting (11):

$$(13) \qquad tc_i^T = tc_i^*(1 + \varepsilon_i + \varepsilon_i \eta_i).$$

Now define g_i to be the reduction in cost attributable to trading:

(14) $\qquad g_i = tc_i^N - tc_i^T = tc_i^*((1 + \varepsilon_i)^{1 + \eta_i} - 1 - \varepsilon_i - \varepsilon_i \eta_i).$

When ε_i is zero, g_i is equal to zero: if abatement is allocated efficiently across countries, there will be no need for trading and costs will be the same whether or not trading is allowed. Moving ε_i away from zero in either direction raises g_i, which can be seen by differentiating (14) with respect to ε_i:

(15) $\qquad \dfrac{dg_i}{d\varepsilon_i} = tc_i^*(1 + \eta_i)((1 + \varepsilon_i)^{\eta_i} - 1).$

The derivative will have the same sign as ε_i, so moving ε_i away from zero in either direction raises the cost advantage of trading.[1] All countries, in other words, will be at least as well off under trading: those with $\varepsilon_i = 0$ will be no worse off and all others will be strictly better off.

What we have shown so far is a result that is very familiar to most economists: allowing emissions permits to be traded can reduce overall abatement costs and will be weakly preferred by all agents to a nontradable system. That statement, however, is purely qualitative and gives no indication about whether the gains are small or large. Moreover, it ignores any administrative or political costs that trading might create. As a result, it is not useful for real-world policy decisions: a trading policy that produced small gains from trade at the cost of large administrative or political problems would be inefficient and counterproductive overall.

One way to move beyond a purely qualitative result is to compare the gains from trade to the value of the international permit transactions necessary to bring them about, since the latter is likely to be closely correlated with the political costs of the policy. Let ϕ_i be the ratio between the two:

(16) $\qquad \phi_i = \dfrac{g_i}{tc_i^P} = \dfrac{(1 + \varepsilon_i)^{1 + \eta_i} - 1 - \varepsilon_i - \varepsilon_i \eta_i}{\varepsilon_i(1 + \eta_i)}.$

---

1. This equation holds when ε_i is greater than -1; if ε_i were less than -1 (if a country was given more than 100 percent of the emissions rights it needed), trading would still be superior but the derivation would be slightly different because total costs in the no-trading case would be zero.

Figure A-1. *Efficiency Gains and Transfers for Different Values of* η

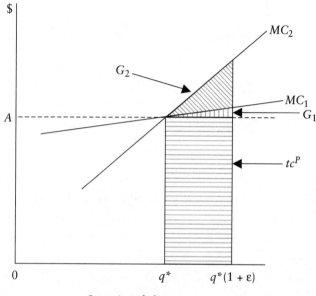

Quantity of abatement

Simplifying this slightly gives:

(17)
$$\phi_i = \frac{(1 + \varepsilon_i)^{1 + \eta_i} - 1}{\varepsilon_i(1 + \eta_i)} - 1.$$

When ϕ_i is large, trading is likely to be a sound, practical policy: the gains from trade are large compared with the transfers it would cause. On the other hand, when ϕ_i is small, it may be that the gains from trade will be overwhelmed by the political costs.

From (17) it is clear that as either ε_i or η_i approaches zero, ϕ_i approaches zero as well. The intuition behind this is straightforward and can be seen from figure A-1, which illustrates the relationship between η, marginal costs, ε and tc^P. Suppose a country is initially allocated fewer than the efficient number of permits and would have to abate $q^*(1 + \varepsilon)$ tons. If the elasticity of marginal abatement costs, η, was small, it would

Table A-1. *Ratio* ϕ *for Alternative Values of* η *and* ε^a

MC elasticity, η	Excess abatement, ε					
	0.05	*0.10*	*0.15*	*0.20*	*0.25*	*0.30*
0.25	0.62	1.22	1.81	2.38	2.95	3.50
0.50	1.24	2.46	3.66	4.84	6.01	7.16
0.75	1.87	3.72	5.56	7.38	9.19	10.99
1.00	2.50	5.00	7.50	10.00	12.50	15.00
1.50	3.78	7.62	11.53	15.49	19.51	23.59
2.00	5.08	10.33	15.75	21.33	27.08	33.00
3.00	7.75	16.03	24.83	34.20	44.14	54.68
4.00	10.51	22.10	34.85	48.83	64.14	80.86

a. Percent.

face a cost curve like MC_1. By buying permits at a total cost shown by area tc^P in the diagram, the country would lower abatement costs enough to produce a net gain equal to area G_1. The net gain is positive but small compared with tc^P because the marginal cost curve is fairly flat. The ratio is larger when the marginal cost curve is steeper: a larger η would lead to a cost curve like MC_2 and a gain of G_2.

The analysis can be taken one step further by evaluating (17) for a range of values of η_i and ε_i. The results are shown in table A-1. If the cost elasticity, η_i, is very high and the permits are allocated very inefficiently (ε_i is large), the efficiency gain will be relatively large compared with the transfer. For example, when $\eta_i = 4$ and $\varepsilon_i = 0.30$, ϕ_i will be about 81 percent. However, if either or both of these parameters are small, the gains from trade will be small compared with the transfer. All of the cells above the gray region, in particular, have gains from trade that are less than 10 percent of the permit transactions. Put another way, in this region the value of permit sales will be more than 10 times the efficiency gains they create.

Frequently Asked Questions

1. Since everything about climate change is so uncertain, wouldn't it be better to wait until more is known before doing anything about it?

There is no uncertainty about the fact that global greenhouse gas concentrations are rising rapidly. Although it is difficult to tell exactly when and how that will affect the climate, no one seriously suggests that it can go on indefinitely with no adverse effects. Because greenhouse gases—especially carbon dioxide—remain in the atmosphere for many years, it is prudent to reduce emissions now where it is possible to do so at low cost.

2. There is no guarantee that permit prices will be equalized across countries. Doesn't that mean that the policy will be inefficient?

Once the policy is in place, permit prices will be equalized at the international trigger price in all countries where emissions exceed the country's allotment of long-term permits. The only circumstance in which prices would not be equalized would be if the emissions of one or more countries were at or below the country's long-term permit allotment. In that case, permit prices would be lower in that country than elsewhere.

The only major country for which this is likely is Russia, whose emissions are currently far below the country's annex B allotment. Because Russia has surplus allowances, however, the difference in permit prices between it and other annex B countries would not reflect differences in the marginal cost of abatement as much as a loosening of the environmental constraint. Permit trading with Russia would lower total abatement costs because total annex B emissions could rise by several hundred million metric tons of carbon, not because of an improvement in the allocation of a fixed amount of abatement.

It is important to note that the Kyoto Protocol does not guarantee that permit prices will be equalized across countries either. Permits originating in different countries will trade at different prices because of variations in the risk of noncompliance across countries. See chapter 4 for details.

In short, our policy does not guarantee that abatement will be allocated efficiently across countries. As a practical matter, however, any inefficiencies are likely to be small. Moreover, the costs of eliminating them through trading may be high; see appendix A for details.

3. An enormous amount of work and years of negotiations have gone into the Kyoto Protocol. Does it really make sense to abandon all that?

In our view, the time and effort devoted to the Kyoto Protocol are the clearest possible evidence that the agreement is fundamentally flawed.

As we argue in chapter 2, climate change is fraught with uncertainty that is unlikely to be resolved for decades. Economic theory, as discussed in chapter 5, provides a very clear foundation for environmental policy under uncertainty, but the Kyoto Protocol is designed in exactly the opposite way.

This protocol's focus on targets and timetables has made negotiations long and difficult because few governments are willing to commit themselves to making very large reductions in their carbon emissions without much idea about either the costs or benefits of doing so.

A realistic, practical policy will have to abandon the targets and timetables approach. It would be better to recognize that as soon as possible by abandoning the protocol instead of spending more years revising it while greenhouse gas emissions grow unchecked.

4. Isn't your proposal just a tax in disguise?

No. Like a tax, our policy will raise the cost of fossil fuels. Unlike with a tax, however, the bulk of the revenue will go to the owners of perpetual permits rather than to the government. The government collects revenue only from the sale of annual permits.

This difference is not just cosmetic. Unlike a tax, our policy creates a constituency with a vested interest in maintaining it: the owners of perpetual permits. As a result, it would be much harder for a government to renege on the policy.

5. Why would developing countries agree to your approach, in which they get nothing from industrialized economies? Under the Kyoto Protocol they will get billions of dollars in payments through the clean development mechanism.

In the long run, full participation by developing countries will be essential to the success of any climate agreement. While the Kyoto Protocol's clean development mechanism provides a limited incentive for partial participation by developing countries in the short run, the protocol's design strongly discourages developing countries from full participation in the long run.

The clean development mechanism is limited by the willingness of developed countries to transfer wealth abroad. At the same time, it has little long-run effect on developing country emissions because it reduces emissions on a project-by-project basis without creating any economy-wide incentives. Moreover, it is by far the most complex of the Kyoto mechanisms, which will make it cumbersome to use. For these reasons, the clean development mechanism is unlikely to transfer much wealth to developing countries or to do much to reduce world emissions.

In addition, the protocol's design makes it very unlikely that developing country participation will ever go much beyond the clean development mechanism. Few developing countries are likely to accede to the full protocol by undertaking annex B commitments. To do so would mean agreeing to meet a rigid constraint on emissions regardless of the consequences for economic growth.

Our proposal, in contrast, is designed to lower the barriers to full participation by developing countries. Because it does not require countries to meet rigid emissions targets, a developing country can join the agreement without the risk that doing so will curtail its growth. In the long run, full participation by developing countries will do far more to control climate change than would the limited incentives provided by the clean development mechanism.

6. Fossil fuel–intensive industries have so far successfully prevented a binding international agreement from being negotiated. Under your approach, governments might actually ratify the agreement. If it was ratified, what would prevent negotiators from subsequently raising the trigger price to an unreasonable level?

The principal safeguard would be each government's desire to avoid unnecessary harm to its economy. In addition, our blueprint contains built-in mechanisms that reduce the incentives for governments or firms to exploit the system. For example, an increase in the trigger price will reduce the number of annual permits sold, potentially lowering the revenue received by each government.

Fossil fuel producers also will face mixed incentives. An increase in the trigger price would reduce the profitability of fuel production, but at the same time it would raise the value of any perpetual permits the firm holds. Because the effects work in opposite directions, firms have far less incentive to lobby against the policy.

7. Why would a country want to enforce your approach domestically? Under the Kyoto Protocol, each country would be forced to comply.

Actually, one of the major stumbling blocks in the negotiations is that a credible enforcement mechanism for the Kyoto Protocol cannot be found. There is no international institution with the power to force countries to comply, nor is there likely to be one in the foreseeable future.

Our proposal, in contrast, is explicitly designed to avoid the need for an international enforcement mechanism. Under our approach, the distribution of perpetual permits to firms and households creates a powerful domestic constituency with an incentive to support the agreement. Permit owners will lobby their own governments to monitor and enforce the policy in order to protect the value of their assets. In addition, governments have an incentive to enforce the policy in order to raise revenue by selling annual permits. Each incentive is internalized within each country.

An added advantage of our approach is that it is more likely to survive noncompliance or even complete withdrawal by some participants. Any given country's defection from the policy has little effect on the remaining participants. Under a global system like the Kyoto Protocol, however, the defection of even one major country would change the world price of emissions permits substantially, causing destabilizing repercussions in all participating countries.

8. Why should fossil fuel producers and users get any perpetual permits? They are the polluters, and usually the polluter should pay.

There are a number of legitimate reasons to give fossil fuel producers and fossil fuel–intensive industries perpetual permits. The first is that the shareholders of these firms will suffer a capital loss from the increase in the price of carbon. The new policy will make previous investments in physical capital—which were made in good faith under the assumption that there would be no drastic changes in the regulation of energy markets—less profitable than expected because the equipment cannot be easily moved to other industries. For a similar reason, workers in energy industries also should be compensated by a permit allocation: many of them will lose their jobs and find that their experience and skills are less valuable in other industries.

In addition, we want to build a constituency for actually putting an effective greenhouse policy in place. Allocating perpetual permits to fossil fuel producers and fossil fuel–intensive industries would create support for the policy.

9. Your approach is too complicated for either politicians or the general public to comprehend. Kyoto is easy to understand.

Kyoto might seem simple, but it has generated thousands of pages of complex text so far and that does not include the additional legislation that will be needed in each participating country to implement the agreement. Because Kyoto is a global system there would need to be significant changes in legal systems around the world in order to establish a uniform system for implementing the protocol.

Our approach, in contrast, takes advantage of the existing legal system and financial markets in each participating country. Perpetual permits will trade in essentially the same way that commodities do now. Because the hybrid policy does not require standardization of the legal systems of participating countries, it is much simpler than Kyoto to implement.

10. Is your approach vulnerable to market manipulation by big players?

One of the major flaws in the Kyoto approach is that countries such as Russia, which are likely to be large suppliers of permits to the global markets, will have market power. Under our approach, however, there is no possibility of market power at the international level because permits trade only within countries. Moreover, market power at the domestic level is severely limited because permit buyers always have the option of buying annual permits from their government at the international trigger price. If a permit owner withholds a permit for strategic reasons, the government will supply an additional permit to the market and the strategy of withholding permits will generate zero return.

11. Some countries might reduce some of their existing energy taxes when adopting the policy so that they can comply with the agreement without really changing anything or reducing their emissions. How would your policy deal with that?

In the short run, there would be nothing to stop countries from doing something along that line. However, it would not have much effect in countries with low energy taxes: existing taxes would not be large enough to prevent the policy from raising the price of greenhouse gas emissions. If the international trigger price was $10, for example, a country with $1's worth of existing taxes gains little by this strategy: it would still need to impose an additional fee of $9.

The countries where this issue would make a difference are those whose energy taxes already are high. A country with $9's worth of existing energy taxes could eliminate them, reducing the net effect of the policy to an increase of only $1. Our policy would not prevent high-tax countries from doing this, but it is important to keep two points in mind. First, countries with high existing energy taxes *already* are helping to reduce greenhouse gas emissions; it might be considered equitable for them to be able to comply with the policy more easily than countries with low taxes. Second, the issue will be unimportant in the long run because a country can rename its existing taxes only once. After that, it will not be able to avoid increasing the price of emissions if the international trigger price is raised in the future. In this respect, it is much like the tariff bindings used in trade negotiations.

12. Are there any unilateral actions that countries should take now, even though international negotiations may not produce a viable agreement in the near future?

Yes—any developed country with a sophisticated financial market should unilaterally issue contingent emissions permits immediately. The role of the permits would depend explicitly on whether a binding international agreement is eventually adopted: with an agreement, a permit would allow the owner to emit a specified carbon-equivalent amount of greenhouse gas; without an agreement, the permit would do nothing. Issuing such permits would allow financial markets to help firms manage the risks posed by potential future climate policies. See chapter 8 for details.

13. What if firms buy permits and go on polluting?

Under the hybrid, firms would have a powerful financial incentive to eliminate any carbon dioxide emissions that can be removed for less than the price of an annual emissions permit. By doing so, a firm could profit by selling its surplus permits or by buying fewer permits than it would otherwise need.

After a point, however, a firm might run out of low-cost abatement opportunities and find that further reductions are very expensive. Once that happens, the firm will stop making additional cuts and will buy permits to cover its remaining emissions. In a sense, it will indeed "buy permits and go on polluting" because its emissions will not have dropped to zero. The key point, however, is that before doing so it will exhaust all of its options to reduce its emissions at a cost less than the price of a permit. Moreover, if the trigger price is set equal to the dollar value of the damages caused by a ton of emissions, the only emissions remaining will be those that cost more to eliminate than the benefits that would be produced are worth.

References

Aldy, Joseph E., Peter R. Orszag, and Joseph E. Stiglitz. 2001. "Climate Change: An Agenda for Global Collective Action." Washington: AEI-Brookings Joint Center for Regulatory Studies (www.aei.brookings.org/publications/related/climate.pdf [August 12, 2002]).

Angell, J. K. "Global, Hemispheric, and Zonal Temperature Deviations Derived from Radiosone Records." *Trends Online: A Compendium of Data on Global Change.* Carbon Dioxide Information Analysis Center, Oak Ridge National Laboratory, U.S. Department of Energy, Oak Ridge, Tenn. For the 2002 version of the data, see http://cdiac.esd.ornl.gov/trends/temp/angell/angell.html [August 27, 2002]).

Arrhenius, Svante. 1896. "On the Influence of Carbonic Acid in the Air upon the Temperature of the Ground." *Philosophical Magazine and Journal of Science* 41.

Cline, William R. 1992. *The Economics of Global Warming.* Washington: Institute for International Economics.

Coase, Ronald H. 1960. "The Problem of Social Cost." *Journal of Law and Economics* 3: 1–44.

Diamond, Peter A., and Jerry A. Hausman. 1994. "Contingent Valuation: Is Some Number Better Than No Number?" *Journal of Economic Perspectives* 8 (4): 45–64.

Energy Information Administration.1999. *International Energy Annual 1999.* Government Printing Office.

————. 2000a. *Annual Energy Review 2000*. Government Printing Office.

————. 2000b. "Emissions of Greenhouse Gases in the United States 1999." Office of Integrated Analysis and Forecasting, U.S. Department of Energy (www.eia.doe.gov/oiaf/1605/gg00rpt [August 12, 2002]).

Energy Journal. 1999. "Special Issue: The Costs of the Kyoto Protocol: A Multi-Model Evaluation."

Hansen, James. 1999. "The Global Warming Debate." Goddard Institute for Space Studies, National Aeronautics and Space Administration (http://www.giss.nasa.gov/edu/gwdebate/ [August 12, 2002])

Intergovernmental Panel on Climate Change. 1990. *Scientific Assessment of Climate Change*. Cambridge University Press.

————. 1995. *Climate Change 1995*. Cambridge University Press.

————. 2000. *Emissions Scenarios*. Cambridge University Press.

————. 2001. *Climate Change 2001: The Scientific Basis: Summary for Policy Makers and Technical Summary of the Working Group I Report*. Cambridge University Press.

————. 2001a. *Climate Change 2001: The Scientific Basis*. Cambridge University Press.

————. 2001b. *Climate Change 2001: Impacts, Adaptation, and Vulnerability*. Cambridge University Press.

————. 2001c. *Climate Change 2001: Mitigation*. Cambridge University Press.

Jiang, Tingsong. 2001. "Economic Instruments of Pollution Control in an Imperfect World: Theory and Implications for Carbon Dioxide Emissions Control in China," Ph.D. dissertation, Australian National University, Canberra.

Jones, P. D., and others. 2000. "Global and Hemispheric Temperature Anomalies—Land and Marine Instrumental Records." *Trends Online: A Compendium of Data on Global Change*. Carbon Dioxide Information Analysis Center, Oak Ridge National Laboratory, U.S. Department of Energy, Oak Ridge, Tenn. (http://cdiac.esd.ornl.gov/trends/temp/jonescru/jones.html [August 12, 2002])

Kopp, Raymond, Richard Morgenstern, and William A. Pizer. 1997. "Something for Everyone: A Climate Policy That Both Environmentalists and Industry Can Live With." *Weathervane* (September 29). Washington: Resources for the Future.

Kopp, Raymond, and others. 1999. "A Proposal for Credible Early Action in U.S. Climate Policy." *Weathervane*. Washington. Resources for the Future. (http://www.weathervane.rff.org/features/feature060.html [August 12, 2002]).

Marland, G., T. A. Boden, and R. J. Andres. 1998 (revised July 2001). "Global, Regional, and National Annual CO2 Emissions from Fossil-Fuel Burning, Cement Production, and Gas Flaring: 1751–1998." NDP-30. Carbon Dioxide Information Analysis Center, Oak Ridge National Laboratory, U.S. Department of Energy, Oak Ridge, Tenn. (http://cdiac.esd.ornl.gov/ndps/ndp030.html [August 12, 2002]).

McKibbin, Warwick J., and Peter J. Wilcoxen. 1997a. "A Better Way to Slow Global Climate Change." *Brookings Policy Brief* 17 (June). Brookings.

———. 1997b. "Salvaging the Kyoto Climate Change Negotiations." *Brookings Policy Brief* 27 (November). Brookings.

McKibbin, Warwick J., Robert Shackleton, and Peter J. Wilcoxen. 1999. "What to Expect from an International System of Tradable Permits for Carbon Emissions." *Resource and Energy Economics* 21 (3/4): 319–46.

Meadows, Donella H., and others. 1972. *The Limits to Growth: A Report for the Club of Rome's Project on the Predicament of Mankind.* New York: Universe Books.

National Academy of Sciences. 2000. *Reconciling Observations of Global Temperature.* Washington: National Academy Press.

National Climate Data Center. 1999. "Climate of 1998: Annual Review." U.S. National Oceanic and Atmospheric Administration.

Newell, Richard G., and William A. Pizer. 1998. "Regulating Stock Externalities under Uncertainty." Discussion Paper 99–10. Washington: Resources for the Future.

Nordhaus, William D. 1991. "The Cost of Slowing Climate Change: A Survey." *Energy Journal* 12 (1).

———. 1992. "The DICE Model: Background and Structure of a Dynamic Integrated Climate-Economy Model of the Economics of Global Warming." Cowles Foundation Discussion Paper 1009. Cowles Foundation for Research in Economics, Yale University.

———. 1993. "Reflections on the Economics of Climate Change." *Journal of Economic Perspectives* 7 (4).

———. 1994. *Managing the Global Commons.* MIT Press.

Nordhaus, William D., and Joseph G. Boyer. 1999. "Requiem for Kyoto." *Energy Journal.* Special Issue: 93–130.

Olivier, J. G. J., and J. J. M. Berdowski. 2001. "Global Emissions Sources and Sinks." In *The Climate System,* edited by J. Berdowski, R. Guicherit, and B. J. Heij, pp. 33–78. Lisse, Netherlands: A. A. Balkema/Swets & Zeitlinger.

Petit, J. R., and others. 2000. "Historical Isotopic Temperature Record from the Vostok Ice Core." *Trends Online: A Compendium of Data on Global Change.* Carbon Dioxide Information Analysis Center, Oak Ridge National Laboratory, U.S. Department of Energy, Oak Ridge, Tenn. (http://cdiac.esd.ornl.gov/trends/temp/vostok/jouz_tem.htm [August 12, 2002]).

Roberts, Marc J., and A. Michael Spence. 1976. "Effluent Charges and Licenses under Uncertainty." *Journal of Public Economics* 5: 193–208.

Rosenzweig, Richard, and others. 2002. *The Emerging International Greenhouse Gas Market.* Arlington, Va.: Pew Center on Global Climate Change.

Stern, D. I., and R. K. Kaufmann.1998. "Annual Estimates of Global Anthropogenic Methane Emissions: 1860–1994." *Trends Online: A Compendium of*

Data on Global Change. Carbon Dioxide Information Analysis Center, Oak Ridge National Laboratory, U.S. Department of Energy, Oak Ridge, Tenn. (http://cdiac.esd.ornl.gov/trends/meth/ch4.htm [August 12, 2002]).

Tol, Richard S. J. 1999. "Kyoto, Efficiency, and Cost-Effectiveness: Applications of FUND," *Energy Journal.* Special Issue: 131–56.

Victor, David. 2001. *The Collapse of the Kyoto Protocol and the Struggle to Slow Global Warming.* Princeton University Press.

Weitzman, Martin L. 1974. "Prices vs. Quantities." *Review of Economic Studies* 41: 477–91.

Index